Somnuek Worawiset

Endomorphism monoids of Strong semilattices of left simple Semigroups

Somnuek Worawiset

Endomorphism monoids of Strong semilattices of left simple Semigroups

Theory in Algebra

Südwestdeutscher Verlag für Hochschulschriften

Impressum / Imprint
Bibliografische Information der Deutschen Nationalbibliothek: Die Deutsche Nationalbibliothek verzeichnet diese Publikation in der Deutschen Nationalbibliografie; detaillierte bibliografische Daten sind im Internet über http://dnb.d-nb.de abrufbar.

Bibliographic information published by the Deutsche Nationalbibliothek: The Deutsche Nationalbibliothek lists this publication in the Deutsche Nationalbibliografie; detailed bibliographic data are available in the Internet at http://dnb.d-nb.de.

Coverbild / Cover image: www.ingimage.com

Verlag / Publisher:
Südwestdeutscher Verlag für Hochschulschriften
ist ein Imprint der / is a trademark of
OmniScriptum GmbH & Co. KG
Heinrich-Böcking-Str. 6-8, 66121 Saarbrücken, Deutschland / Germany
Email: info@svh-verlag.de

Herstellung: siehe letzte Seite /
Printed at: see last page
ISBN: 978-3-8381-5046-8

Zugl. / Approved by: Oldenburg, 2011

Acknowledgements

I would like to thank my supervisor Professor Dr. Dr. h.c. Ulrich Knauer for his advice and enthusiasm. Without his assistance and encouragement, especially in the suggestion of possible results, and his demands for clarity in expression, it is doubtful whether this thesis would have been completed.

In addition I am grateful to the Department of Mathematics, Carl von Ossietzky Universität of Oldenburg, Germany who helped sustain me during my years of research.

I would like to thank Professor Norman R. Reilly and Professor Friedrich Wehrung for suggestions many examples and get some ideas for this thesis.

I also would like to thank Khon Kaen University, Thailand and the Royal Thai Government for the support of my research.

Finally I would like to thank my friends, my family, especially my parents. Without their support this thesis would never have been started, nor completed.

Somnuek Worawiset

Abstract

Endomorphism monoids have long been of interest in universal algebra and also in the study of particular classes of algebraic structures.

For any algebra, the set of endomorphisms is closed under composition and forms a monoid (that is, a semigroup with identity). The endomorphism monoid is an interesting structure from a given algebra.

In this thesis we study the structure and properties of the endomorphism monoid of a strong semilattice of left simple semigroups. In such semigroup we consider mainly that the defining homomorphisms are constant or isomorphisms. For arbitrary defining homomorphisms the situation is in general extremely complicated, we have discussed some of the problems at the end of the thesis.

First we consider conditions, under which the endomorphism monoids are regular, idempotent-closed, orthodox, left inverse, completely regular and idempotent.

Later, as corollaries we obtain results for strong semilattices of groups which are known under the name of Clifford semigroups and we also consider strong semilattices of left or right groups as well. Both are special cases of the strong semilattices of left simple semigroups.

Abstract

Endomorphismenmonoide schon lange von Interesse in der universellen Algebra und werden für die Untersuchung bestimmter Klassen von algebraischen Strukturen eingesetzt.

Für jede Algebra ist die Menge der Endomorphismen abgeschlossen unter Komposition und bildet ein Monoid (das heißt, eine Halbgruppe mit einem neutralen Element).

In dieser Arbeit untersuchen wir die Strukturen und Eigenschaften des Endomorphismenmonoids eines starken Halbverbands von links einfachen Halbgruppen. Für solche Halbgruppe betrachten wir vor allem die Situation, dass die definierenden Homomorphismen konstant sind oder Isomorphismen. Für beliebige definierende Homomorphismen ist die Lage im Allgemeinen äußerst kompliziert, wir haben einige von ihnen diskutiert, aber es bleiben viele offene Probleme.

Zunächst untersuchen wir die Bedingungen, unter denen die Endomorphismenmonoide regulär, idempotent abgeschlossen, orthodox, linksinvers, vollständig regulär oder idempotent sind.

Später erhalten wir als Folgerungen die entsprechenden Ergebnisse für starke Halbverbände von Gruppen, die unter dem Namen Clifford Halbgruppen bekannt sind, und ebenso für starke Halbverbände von Links- oder Rechtsgruppen. Alles sind Sonderfälle der starken Halbverbände von links einfachen Halbgruppen.

Summary

Endomorphism monoids have long been of interest in universal algebra and also in the study of particular classes of algebraic structures.

For any algebra, the set of endomorphisms is closed under composition and forms a monoid (that is, a semigroup with identity). The endomorphism monoid is an interesting structure from a given algebra.

In this thesis we study the structure and properties of the endomorphism monoid of a strong semilattice of left simple semigroups. In such semigroup we consider mainly that the defining homomorphisms are constant or bijective. For arbitrary defining homomorphisms the situation is in general extremely complicated, we have discussed some of the problems at the end of each chapters. The semigroups, which are considered are finite.

Let Y be a semilattice and let S_ξ be a semigroup for each $\xi \in Y$ with $S_\alpha \cap S_\beta = \emptyset$ if $\alpha \neq \beta, \alpha, \beta \in Y$. For each pair $\alpha, \beta \in Y$ with $\alpha \geq \beta$, let $\varphi_{\alpha,\beta} : S_\alpha \to S_\beta$ be a semigroup homomorphism such that $\varphi_{\alpha,\alpha}$ is the identity mapping, and if $\alpha > \beta > \gamma$ then $\varphi_{\alpha,\gamma} = \varphi_{\beta,\gamma}\varphi_{\alpha,\beta}$.

Consider $S = \bigcup\limits_{\alpha \in Y} S_\alpha$ with multiplication

$$a * b = \varphi_{\alpha,\alpha\wedge\beta}(a)\varphi_{\beta,\alpha\wedge\beta}(b)$$

for $a \in S_\alpha$ and $b \in S_\beta$. The semigroup S is called a *strong semilattice Y of semigroups* S_ξ. For $\alpha, \beta \in Y$ we call $\varphi_{\alpha,\beta}$ the *defining homomorphisms* of S also called *structure homomorphisms*. We denote a strong semilattice of semigroups S_α with defining homomorphisms $\varphi_{\alpha,\beta}$ by $S = [Y; S_\alpha, \varphi_{\alpha,\beta}]$. A *strong semilattice of groups* is known under the name a *Clifford semigroup.*

Since all S_α is finite, we denote an idempotent e_α as a fixed element with corresponding to S_α.

Now we study the endomorphism monoids of the strong semilattices of left simple semigroups with constant defining homomorphisms c_{α,e_β} and denoted by $S = [Y; S_\alpha, e_\alpha, c_{\alpha,e_\beta}]$.

We obtain the following results:

Theorem: Let $S = [Y; S_\alpha, e_\alpha, c_{\alpha,e_\beta}]$ be a non-trivial strong semilattice of left simple semigroups with $\nu = \wedge Y$.

If the monoid $End(S)$ is **regular** then the following conditions hold

1) the monoid $End(Y)$ is regular,

2) the set $Hom(S_\nu, S_\alpha)$ consists of constant

mappings for all $\alpha \in Y$ with $\nu < \alpha$, and

3) the set $Hom(S_\alpha, S_\beta)$ is hom-regular for every $\alpha, \beta \in Y$.

If the following conditions hold

1) $Y = Y_{0,n}$,

2) the set $Hom(S_0, S_\alpha)$ consists of constant mappings for all $\alpha \in Y$ with $\alpha \neq 0$,

3) the set $Hom(S_\alpha, S_\beta)$ is hom-regular for every $\alpha, \beta \in Y_{0,n}$,

4) S_0 contains one idempotent e_0,

then the monoid $End(S)$ is regular.

The monoid $End(S)$ is **idempotent-closed** if and only if

1) $Y = Y_{0,n}$ and

2) the monoid $End(S_\xi)$ is idempotent-closed for every $\xi \in Y_{0,n}$.

If monoid $End(S)$ is **orthodox** then the following conditions hold

1) $Y = Y_{0,n}$,

2) the set $Hom(S_0, S_\alpha)$ consists of constant mappings for all $\alpha \in Y_{0,n}$ with $\alpha \neq 0$,

3) the monoid $End(S_\xi)$ is idempotent-closed for all $\xi \in Y_{0,n}$, and

4) the set $Hom(S_\alpha, S_\beta)$ is hom-regular for every $\alpha, \beta \in Y_{0,n}$.

If the following conditions hold

1) $Y = Y_{0,n}$,

2) the set $Hom(S_0, S_\alpha)$ consists of constant mappings for all $\alpha \in Y_{0,n}$, $\alpha \neq 0$,

3) the monoid $End(S_\xi)$ is idempotent-closed for all $\xi \in Y_{0,n}$,

4) the set $Hom(S_\alpha, S_\beta)$ is hom-regular for every $\xi \in Y_{0,n}$, and

5) S_0 contains one idempotent e_0,

then the monoid $End(S)$ is orthodox.

Then the monoid $End(S)$ is **left inverse** if and only if

1) $Y = Y_{0,n}$ and

2)the monoid $End(S_\xi)$ is left inverse for every $\xi \in Y_{0,n}$.

If the monoid $End(S)$ is **completely regular** then the following conditions hold

1) $|Y| = 2$,

2) the set $Hom(S_\nu, S_\alpha)$ consists of constant

mappings for all $\alpha \in Y$ with $\nu < \alpha$, and

3) the monoid $End(S_\xi)$ is completely regular for every $\xi \in Y$.

If the following conditions hold

1) $|Y| = 2$,

2) the set $Hom(S_\nu, S_\alpha)$ consists of constant

mappings for all $\alpha \in Y$, $\alpha \neq \nu$,

3) the monoid $End(S_\xi)$ is completely regular

for every $\xi \in Y$, and

4) S_ν has only one idempotent,

then the monoid $End(S)$ is completely regular.

Then the monoid $End(S)$ is **idempotent** if and only if

1) $|Y| = 2$,

2) the set $Hom(S_\nu, S_\alpha)$ consists of constant

mappings for all $\alpha \in Y$ with $\nu < \alpha$, and

3) the monoid $End(S_\xi)$ is idempotent for every $\xi \in Y$.

Moreover, if the defining homomorphisms $\varphi_{\alpha,\beta}$ are bijective for all $\alpha, \beta \in Y$ such that $\beta < \alpha$, we found that:

Let $S = [Y; T_\alpha, e_\alpha, \varphi_{\alpha,\beta}]$, $T_\xi \cong T$ be a non-trivial strong semilattice of left simple semigroups. Then the monoid $End(S)$ is regular (idempotent-closed, orthodox, left inverse, completely regular, and idempotent) if and only if the monoids $End(Y)$ and $End(T)$ have such property.

We also consider the endomorphism monoids of the strong semilattices of left simple semigroups with surjective defining homomorphisms $\varphi_{\alpha,\beta}$ and $Y = Y_{0,n}$.

Let $Y = Y_{0,n}$ and let $S = [Y_{0,n}; S_\alpha, e_\alpha, \varphi_{\alpha,\beta}]$ be a non-trivial strong semilattice of left simple semigroup with surjective defining homomorphisms $\varphi_{\alpha,\beta}$.

Then the monoid $End(S)$ is **regular** if and only if the set $Hom(S_\alpha, S_{\beta)})$ is hom-regular for all $\alpha, \beta \in Y_{0,n}$.

Then the monoid $End(S)$ is **idempotent-closed** if and only if the monoid $End(S_\xi)$ is idempotent-closed for every $\xi \in Y_{0,n}$.

Then the monoid $End(S)$ is **orthodox** if and only if the following conditions hold

1) the monoid $End(S_\xi)$ is idempotent-closed

for every $\xi \in Y_{0,n}$, and

2) the set $Hom(S_\alpha, S_\beta)$ is hom-regular

for every $\alpha, \beta \in Y_{0,n}$.

Then the monoid $End(S)$ is **left inverse** if and only if the monoid $End(S_\xi)$ is left inverse for every $\xi \in Y_{0,n}$.

Then the monoid $End(S)$ is **completely regular** if and only if $|Y| = 2$ and the monoid $End(S_\xi)$ is completely regular for every $\xi \in Y$.

Then the monoid $End(S)$ is **idempotent** if and only if $|Y| = 2$ and the monoid $End(S_\xi)$ is idempotent for each $\xi \in Y$.

Zusammenfassung

Endomorphismenmonoide sind schon lange von Interesse in der universellen Algebra und werden für die Untersuchung bestimmter Klassen von algebraischen Strukturen eingesetzt.

Für jede Algebra ist die Menge der Endomorphismen abgeschlossen unter Komposition und bildet ein Monoid (das heißt, eine Halbgruppe mit einem neutralen Element).

In dieser Arbeit untersuchen wir die Strukturen und Eigenschaften des Endomorphismenmonoids eines starken Halbverbands von links einfachen Halbgruppen. Für solche Halbgruppe betrachten wir vor allem die Situation, dass die definierenden Homomorphismen konstant sind oder Isomorphismen. Für beliebige definierende Homomorphismen ist die Lage im Allgemeinen äußerst kompliziert, wir haben einige von ihnen diskutiert, aber es bleiben viele offene Probleme. Die Halbgruppen, die angesehen werden, sind endlich.

Sei Y ein Halbverband und sei S_ξ eine Halbgruppe für jedes $\xi \in Y$ mit $S_\alpha \cap S_\beta = \emptyset$, wenn $\alpha \neq \beta$, $\alpha, \beta \in Y$. Für jedes Paar $\alpha, \beta \in Y$ mit $\alpha \geq \beta$, sei $\varphi_{\alpha,\beta} : S_\alpha \to S_\beta$ ein Halbgruppen Homomorphismus, so dass $\varphi_{\alpha,\alpha}$ die Identitätsabbildung ist, und wenn $\alpha > \beta > \gamma$, dann $\varphi_{\alpha,\gamma} = \varphi_{\beta,\gamma}\varphi_{\alpha,\beta}$.

Wir betrachten $S = \bigcup\limits_{\alpha \in Y} S_\alpha$ mit Multiplikation

$$a * b = \varphi_{\alpha,\alpha\wedge\beta}(a)\varphi_{\beta,\alpha\wedge\beta}(b)$$

für $a \in S_\alpha$ und $b \in S_\beta$. Die Halbgruppe S wird als *stark Halbverband Y von Halbgruppen S_ξ* bezeichnet. Für $\alpha, \beta \in Y$ nennen wir $\varphi_{\alpha,\beta}$ die *definierenden Homomorphismen* von S oder auch *Struktur Homomorphismen.* Wir bezeichnen einen starken Halbverband von Halbgruppen S_α mit definierenden Homomorphismen $\varphi_{\alpha,\beta}$ durch $S = [Y; S_\alpha, \varphi_{\alpha,\beta}]$. Ein *starker Halbverband von Gruppen* ist bekannt unter dem Namen *Clifford Halbgruppe.*

Da alle S_α endlich sind, wählen wir eines der Idempotenten $e_\alpha \in S_\alpha$ aus.

Jetzt studieren wir die Endomorphismenmonoide der starken Halbverbände von links einfachen Halbgruppen $S = [Y; S_\alpha, e_\alpha, c_{\alpha,e_\beta}]$ mit konstanten definierenden Homomorphismen c_{α,e_β}.

Wir erhalten die folgenden Ergebnisse:

Sei $S = [Y; S_\alpha, e_\alpha, c_{\alpha,e_\beta}]$ ein nicht-trivialer starker Halbverband von links einfachen Halbgruppen mit $\nu = \wedge Y$.

Wenn das Monoid $End(S)$ **regulär** ist, dann gelten die folgenden Bedingungen

1) das Monoid *End*(Y) ist regulär,

2) die Menge $Hom(S_\nu, S_\alpha)$ besteht aus konstanten Abbildungen für alle $\alpha \in Y$ mit $\nu < \alpha$ und

3) die Menge $Hom(S_\alpha, S_\beta)$ ist hom-regulär für alle $\alpha, \beta \in Y$.

Wenn die folgenden Bedingungen erfüllt sind

1) $Y = Y_{0,n}$,

2) die Menge $Hom(S_\nu, S_\alpha)$ besteht aus konstanten Abbildungen für alle $\alpha \in Y_{0,n}$ mit $\alpha \neq 0$,

3) die Menge $Hom(S_\alpha, S_\beta)$ ist hom-regulär für alle $\xi \in Y_{0,n}$, und

4) S_0 enthält nur ein idempotentes Element,

dann ist das Monoid *End*(S) regulär.

Das Monoid *End*(S) ist **idempotent-abgeschlossen** genau dann, wenn

1) $Y = Y_{0,n}$ und

2) das Monoid *End*(S_ξ) ist idempotent-abgeschlossen für jedes $\xi \in Y_{0,n}$.

Das Monoid *End*(S) ist **orthodox** genau dann, wenn

1) $Y = Y_{0,n}$,

2) die Menge $Hom(S_0, S_\alpha)$ besteht aus konstanten Abbildungen für alle $\alpha \in Y_{0,n}$ mit $\alpha \neq 0$,

3) das Monoid *End*(S_ξ) ist idempotent-abgeschlossen für alle $\xi \in Y_{0,n}$, und

4) die Menge $Hom(S_\alpha, S_\beta)$ ist hom-regulär für alle $\alpha, \beta \in Y_{0,n}$.

Wenn die folgenden Bedingungen erfüllt sind

1) $Y = Y_{0,n}$,

2) die Menge $Hom(S_0, S_\alpha)$ besteht aus konstanten Abbildungen für alle $\alpha \in Y_{0,n}$ mit $\alpha \neq 0$,

3) das Monoid *End*(S_ξ) ist idempotent-abgeschlossen für alle $\xi \in Y_{0,n}$,

4) die Menge $Hom(S_\alpha, S_\beta)$ ist hom-regulär für alle $\xi \in Y_{0,n}$, und

5) S_0 enthält nur ein idempotentes Element,

dann ist das Monoid *End*(S) orthodox.

Das Monoid *End*(S) ist **linksinvers** genau dann, wenn

1) $Y = Y_{0,n}$ und

2) das Monoid $End(S_\xi)$ ist linksinvers

für alle $\xi \in Y_{0,n}$.

Das Monoid $End(S)$ ist **vollständig regulär** wenn die folgenden Bedingungen erfüllt sind

1) $|Y| = 2$,

2) die Menge $Hom(S_\nu, S_\alpha)$ besteht aus

konstanten Abbildungen für alle $\alpha \in Y$ mit $\nu < \alpha$, und

3) das Monoid $End(S_\xi)$ ist vollständig regulär für alle $\xi \in Y$.

Wenn die folgenden Bedingungen erfüllt sind

1) $|Y| = 2$,

2) die Menge $Hom(S_\nu, S_\alpha)$ besteht aus

konstanten Abbildungen für alle $\alpha \in Y$ mit $\nu < \alpha$,

3) das Monoid $End(S_\xi)$ ist vollständig regulär für alle $\xi \in Y$, und

4) S_ν enthält nur ein idempotentes Element,

dann ist das Monoid $End(S)$ vollständig regulär.

Das Monoid $End(S)$ ist **idempotent** genau dann, wenn

1) $|Y| = 2$,

2) die Menge $Hom(S_\nu, S_\alpha)$ besteht aus

konstanten Abbildungen für alle $\alpha \in Y$ mit $\nu < \alpha$, und

3) die Monoid $End(S_\xi)$ ist idempotent für alle $\xi \in Y$.

Außerdem, wenn $\varphi_{\alpha,\beta}$ für alle $\alpha, \beta \in Y$ bijektiv ist, so dass $\beta < \alpha$, finden wir:

Sei $S = [Y; T_\alpha, e_\alpha, \varphi_{\alpha,\beta}]$, $T_\xi \cong T$ ein nicht-trivialer starker Halbverband von links einfachen Halbgruppen. Dann ist das Monoid $End(S)$ regulär (idempotent-abgeschlossen, orthodox, linksinvers, vollständig regulär, und idempotent) genau dann, wenn die Monoiden $End(Y)$ und $End(T)$ eine solche Eigenschaft haben.

Wir betrachten auch die Endomorphismenmonoide der starken Halverband von links einfachen Halbgruppen mit surjektiv definierenden Homomorphismen $\varphi_{\alpha,\beta}$ und $Y = Y_{0,n}$.

Sei $Y = Y_{0,n}$ und sei $S = [Y_{0,n}; S_\alpha, e_\alpha, \varphi_{\alpha,\beta}]$ ein nicht-trivialer starker Halbverband von links einfachen Halbgruppen mit surjektiv definierenden Homomorphismen $\varphi_{\alpha,\beta}$.

Dann ist das Monoid $End(S)$ **regulär** genau dann, wenn die eingestellte $Hom(S_\alpha, S_\beta)$ hom-regulär ist für alle $\alpha, \beta \in Y_{0,n}$.

Das Monoid $End(S)$ ist **idempotent-abgeschlossen** genau dann, wenn das Monoid $End(S_\xi)$ idempotent-abgeschlossen ist für jedes $\xi \in Y_{0,n}$.

Das Monoid $End(S)$ ist **orthodox** genau dann, wenn folgende Bedingungen erfüllt sind

1) Das Monoid $End(S_\xi)$ ist idempotent-abgeschlossen
   für jedes $\xi \in Y_{0,n}$ und
2) Die Menge $Hom(S_\alpha, S_\beta)$ ist hom-regulär
   für jedes $\alpha, \beta \in Y_{0,n}$.

Das Monoid $End(S)$ ist **linksinvers** genau dann, wenn das Monoid $End(S_\xi)$ linksinvers ist für jedes $\xi \in Y_{0,n}$.

Das Monoid $End(S)$ ist **vollständig regulär** genau dann, wenn $|Y| = 2$ und das Monoid $End(S_\xi)$ vollständig regulär ist für jedes $\xi \in Y$.

Das Monoid $End(S)$ ist **idempotent** genau dann, wenn $|Y| = 2$ und das Monoid $End(S_\xi)$ idempotent ist für jedes $\xi \in Y$.

Contents

Introduction

Endomorphism monoids have long been of interest in universal algebra and also in the study of particular classes of algebraic structures.

For any algebra, the set of endomorphisms is closed under composition and forms a monoid (that is, a semigroup with identity). The endomorphism monoid is an interesting structure from a given algebra. Some properties have been investigated, regular for example. For many algebras the endomorphism monids have been studied, for example, in [3], posets whose monoids of order-preserving maps are regular, the regularity and substructures of Hom of modules have been in [6]. The endomorphism monoids of some special groups were studied. Endomorphism rings of abelian groups have been studied in [15] and endomorphism monoids of the generalized quaternion groups, dihedral 2-groups, the alternating group A_4 and symmetric groups were considered by Puusemp[16], [18], [19]. In [20] the endomorphisms of Clifford semigroups were described.

In this thesis we study properties of the endomorphism monoids of strong semilattices of left simple semigroups; namely regular endomorphisms, idempotent-closed sets of endomorphisms, orthodox sets of endomorphisms, left inverse endomorphisms, completely regular endomorphisms and idempotent endomorphisms.

This thesis contains 7 chapters; Chapter 1 is of an introductory nature which provides basic definitions and reviews some of the background material which is needed for reading the subsequent chapters. We also introduce the concept of homomorphism regularity of two groups.

In Chapter 2 we mentioned finite semilattices whose endomorphism monoids are regular and we investigated the above regularity properties of the endomorphism monoids of finite semilattices and of sets.

In Chapter 3 we consider strong semilattices of left simple semigroups whose endomorphism monoids have the above regularity properties. In this chapter we consider strong semilattices of left simple semigroups in which the defining homomorphisms are constant or isomorphisms.

The results in this chapter are valid for the endomorphism monoids of strong semilattices of right simple semigroups as well.

In Chapter 4 we consider Clifford semigroups, i.e., strong semilattices of groups whose endomorphism monoids have the above regularity properties. In this chapter we consider Clifford semigroups in which the defining homomorphisms are constant or bijective.

In Chapter 5 we consider strong semilattices of left groups whose endomorphism monoids have the above regularity properties. In this chapter we consider strong semilattices of left groups in which the defining homomorphisms are constant or bijective.

All results in Chapter 4 and Chapter 5 are as a consequence of Chapter 2.

In Chapter 6 we consider strong semilattices of left simple semigroups whose endomorphism monoids have the above regularity properties. In this chapter we consider the strong semilattice of left simple semigroups in which the defining homomorphisms are surjective with the semilattice $Y_{0,n}$.

In Chapter 7 we discuss the strong semilattices of left simple semigroups with a two-element chain in which the defining homomorphisms are arbitrary.

Symbols

Symbol	Description	Page
$E(S)$	the set of idempotents of S	5
$\varphi_{\alpha,\beta}$	defining homomorphisms	6
$Hom(G, H)$	homomorphisms from G to H	6
$End(G)$	endomorphism monoid of G	6
n, m	positive integers	6
$\mathbb{Z}_n$	group modulo n	6
$V(a)$	the set of inverses of a	10
c_x	the constant map onto x	11
$G = A \ltimes B$	G is a normal direct sum of A by B	13
$Im(f)$	image under f	13
$Ker(f)$	kernel of f	13
$A \oplus B$	A is a direct sum with B	14
$\mathbb{Z}_n$	the cyclic group of order n	16
$\mathbb{Z}_p$	group modulo a prime p	16
Q	the Quaternion group	16
D_n	the dihedral group D_n	16
0	the minimal element of $Y_{0,n}$	21
$K_{1,n}$	a bipartite graph	21
$Y_{0,n}$	a semilattice with minimum 0	21
$[Y, S_\alpha, e_\alpha, \varphi_{\alpha,\beta}]$	a strong semilattaice of left simple semigroups	25
ν	the minimal element of a semilattice Y	35
μ	the maximal element of a semilattice Y	35
$L_n \times G$	a left group	61
$[Y, L_{n_\xi} \times G_{n_\xi}, \varphi_{\alpha,\beta}]$	a strong semilattaice of left groups	61
$[Y, G_\alpha, \varphi_{\alpha,\beta}]$	a Clifford semigroup	71

Chapter 1

Preliminaries

In this chapter we shall provide some basic knowledge of semigroup theory that will be used in this thesis. Section 1.2 contains some considerations on ordered Clifford semigroups showing that all endomorphisms preserve order. So the original idea to study order preserving endomorphisms of Clifford semigroups does not lead anywhere.

Section 1.3 develops some new aspects in the study of homomorphism sets $Hom(G, H)$ where G and H are groups. With a slight generalization of endomorphism we present the concept of hom-regularity. This section is based on semigroup theory which are found in [5], [13] and [14].

1.1 General background for semigroups

Definition 1.1.1. Let X be a nonempty set. A *binary relation* ρ on X is a subset of the cartesian product $X \times X$; for membership in ρ, we write $x\rho y$ but occasionally also $(x, y) \in \rho$.

A semigroup is an algebraic structure consisting of a nonempty set S together with an associative binary operation.

Definition 1.1.2. A semigroup S is called *commutative* if $ab = ba$ for any $a, b \in S$. An element $a \in S$ is called *idempotent* if $a^2 = a$. Denote by $E(S)$ the set of all idempotents of a semigroup S.

Definition 1.1.3. A (meet)-semilattice $(S, \wedge)$ is a commutative semigroup in which each element is idempotent. A partial ordering is defined on S by $a \leq b$ if and only if $a \wedge b = a$, with respect to this order, each pair of elements of S has a greatest lower bound, or *meet*, which coincides with the operation $\wedge$. If each pair of elements of S also has a least upper bound, or *join* (denoted $\vee$), then S is said to be a *lattice*.

Definition 1.1.4. Let S and T be semigroups and let $f : S \to T$ be a mapping, then f is called a *semigroup homomorphism* if $f(xy) = f(x)f(y)$ for all $x, y \in S$. The set of semigroup homomorphisms is denoted by $Hom(S,T)$ and $End(S) = Hom(S,S)$. The set $End(S)$ forms a monoid with composition as a multiplicative and mappings are composed from right to left. For $f, g \in End(S)$, the composition of f and g is written as $g \circ f$ and $(g \circ f)(x) = g(f(x))$ for all $x \in S$, we write gf instead of $g \circ f$.

Definition 1.1.5. Let Y be a meet-semilattice and let S_ξ be a semigroup for each $\xi \in Y$ with $S_\alpha \cap S_\beta = \emptyset$ if $\alpha \neq \beta, \alpha, \beta \in Y$. For each pair $\alpha, \beta \in Y$ with $\alpha \geq \beta$, let $\varphi_{\alpha,\beta} : S_\alpha \to S_\beta$ be a semigroup homomorphism such that $\varphi_{\alpha,\alpha}$ is the identity mapping, and if $\alpha > \beta > \gamma$ then $\varphi_{\alpha,\gamma} = \varphi_{\beta,\gamma}\varphi_{\alpha,\beta}$.

Consider $S = \bigcup_{\alpha \in Y} S_\alpha$ with multiplication

$$a * b = \varphi_{\alpha,\alpha\wedge\beta}(a)\varphi_{\beta,\alpha\wedge\beta}(b)$$

for $a \in S_\alpha$ and $b \in S_\beta$. The semigroup S is called a *strong semilattice Y of semigroups S_ξ*. For $\alpha, \beta \in Y$ we call $\varphi_{\alpha,\beta}$ the *defining homomorphisms* of S also called *structure homomorphisms*, for example [14]. We denote a strong semilattice of semigroups S_α with defining homomorphisms $\varphi_{\alpha,\beta}$ by $S = [Y; S_\alpha, \varphi_{\alpha,\beta}]$. If we replace 'semigroup' by 'group', we call *a strong semilattice of groups* [14] which is known under the name *Clifford semigroup.*

From now on we write $\alpha\beta$ instead of $\alpha \wedge \beta$.

Definition 1.1.6. A semigroup S is called *idempotent-closed*, if

$$\text{for all } a, b \in S \text{ and } a^2 = a,\ b^2 = b,\ \text{one has } (ab)^2 = (ab).$$

Example 1.1.1. If T is a commutative semigroup, then T is idempotent-closed because for idempotents $a, b \in T$ we have $(ab)^2 = abab = a^2b^2 = ab$. For example, the monoid $End(\mathbb{Z}_4)$ is idempotent-closed since $End(\mathbb{Z}_4, \circ) \cong (\mathbb{Z}_4, \cdot)$ (see [9]) is a commutative semigroup.

Example 1.1.2. The monoid $End(\mathbb{Z}_2 \times \mathbb{Z}_2)$ is not idempotent-closed. To see this, take idempotents $f, g \in End(\mathbb{Z}_2 \times \mathbb{Z}_2)$ as follows.

$$f = \begin{pmatrix} 00 & 01 & 10 & 11 \\ 00 & 01 & 00 & 01 \end{pmatrix}$$

and

$$g = \begin{pmatrix} 00 & 01 & 10 & 11 \\ 00 & 10 & 10 & 00 \end{pmatrix}$$

but

$$gf = \begin{pmatrix} 00 & 01 & 10 & 11 \\ 00 & 10 & 00 & 10 \end{pmatrix}$$

is not idempotent.

Definition 1.1.7. An element a of a semigroup S is called *regular* if there exists an element $x \in S$ such that $a = axa$ and if $ax = xa$, then a is called *completely regular.* The semigroup S is called *regular* if all its elements are regular and S is called *completely regular* if all its elements are completely regular. A regular semigroup which is idempotent-closed is called *orthodox.* A semigroup is called a Clifford semigroup if it is completely regular and its idempotents commute. A semigroup S is called *left inverse* if for any idempotents $a, b \in S,\ aba = ab$. A semigroup S is called *right inverse* if for any idempotents $a, b \in S,\ aba = ba$.

According to [13], Petrich formulated the fundamental theorem for the global structure of completely regular semigroups as follows.

Theorem 1.1.1. *Let S be a semigroup. Then the following statements are equivalent:*

(1) S is completely regular.

(2) S is a union of (disjoint) groups.

(3) For every $a \in S,\ a \in aSa^2$.

Remark 1.1.1. We observe that

idempotent $\Rightarrow$ completely regular $\Rightarrow$ regular,

idempotent $\Rightarrow$ idempotent-closed,

commutative $\Rightarrow$ idempotent-closed,

group $\Rightarrow$ inverse $\Rightarrow$ left inverse (right inverse).

In some papers of Puusemp, for example in [17], idempotents of the endomorphism monoids of groups are investigated. That is for any group G, and for each idempotent $f \in End(G)$. Then G can be expressed as a direct product of $Im(f)$ and $Ker(f)$.

The following theorem gives several alternative definitions of a Clifford semigroup which can be found in [5] Theorem 4.2.1.

Theorem 1.1.2. *Let S be a semigroup. Then the following statements are equivalent:*

(1) S is a Clifford semigroup,

(2) S is a semilattice of groups,

(3) S is a strong semilattice of groups,

(4) S is regular, and the idempotents of S are central.

In [13] all homomorphisms of Clifford semigroups are described. The following theorem is Proposition II.2.8 of [13].

Theorem 1.1.3. *Let $S = [Y; G_\alpha, \varphi_{\alpha,\beta}]$ and $T = [Z; H_\alpha, \psi_{\alpha,\beta}]$ be Clifford semigroups. Let $\eta : Y \to Z$ be a homomorphism, for each $\alpha \in Y$, let $f_\alpha : G_\alpha \to H_{\eta(\alpha)}$ be a homomorphism, and assume that for any $\alpha \geq \beta$, the diagram*

$$\begin{array}{ccc} G_\alpha & \overset{f_\alpha}{\rightarrow} & H_{\eta(\alpha)} \\ \varphi_{\alpha,\beta}\downarrow & & \downarrow\psi_{\eta(\alpha),\eta(\beta)} \\ G_\beta & \overset{f_\beta}{\rightarrow} & H_{\eta(\beta)} \end{array}$$

commutes. Define a mapping f on S by $f(a) := f_\alpha(a)$ if $a \in G_\alpha$. Then f is a homomorphism of S into T. Moreover, f is one-to-one (respectively a bijection) if and only if η and all f_α are one-to-one (respectively bijections). Conversely, every homomorphism of S into T can be constructed this way.

In [20] Lemma 1.3 has also described all homomorphisms of two Clifford semigroups which are shown as follows.

Lemma 1.1.1. *Let $S = [Y; G_\alpha, \varphi_{\alpha,\beta}]$ and $T = [Z; H_\alpha, \psi_{\alpha,\beta}]$ be Clifford semigroups. Given a semilattice homomorphism $f_L : Y \to Z$ and a family of group homomorphisms $\{f_\alpha \in Hom(G_\alpha, H_{f_L(\alpha)}) \mid \alpha \in Y\}$ satisfies*

$$f_\beta\varphi_{\alpha,\beta} = \psi_{f_L(\alpha),f_L(\beta)}f_\alpha,$$

for all $\alpha, \beta \in Y$. Then $f : S \to T$ defined by $f(x_\alpha) := f_\alpha(x_\alpha)$ for every $x_\alpha \in S$, $\alpha \in Y$ is a homomorphism of semigroups.

Corollary 1.1.1. *Let* $S = [Y; G_\alpha, \varphi_{\alpha,\beta}]$ *and* $T = [Z; H_\alpha, \psi_{\alpha,\beta}]$ *be Clifford semigroups. Let* $f : S \to T$ *be a homomorphism with the set* $\{f_\alpha \in Hom(S_\alpha, T_{\underline{f}(\alpha)}) \mid \alpha \in Y\}$ *of family of semigroup homomorphisms. If* $\alpha, \beta \in Y, \beta \leq \alpha$ *then*

$$f_\beta(Im(\varphi_{\alpha,\beta})) \subseteq Im(\psi_{\underline{f}(\alpha),\underline{f}(\beta)})$$

$$f_\alpha(Ker(\varphi_{\alpha,\beta})) \subseteq Ker(\psi_{\underline{f}(\alpha),\underline{f}(\beta)}).$$

Lemma 1.1.2. *Let* $S = [Y; G_\alpha, \varphi_{\alpha,\beta}]$ *be a Clifford semigroup. Let* $f, g \in End(S)$. *Write* $h = gf$. *Then*

$$h_\alpha = g_{\underline{f}(\alpha)} f_\alpha$$

for all $\alpha \in Y$.

1.2 Partial orders on Clifford semigroups

In this section we study partial orders on Clifford semigroups and find that all endomorphisms are order-preserving.

Definition 1.2.1. A binary relation ρ on X is

reflexive if $x\rho x$, for all $x \in X$,

symmetric if $x\rho y$ implies that $y\rho x$,

antisymmetric if $x\rho y$ and $y\rho x$ imply that $x = y$, and

transitive if $x\rho y$ and $y\rho z$ imply that $x\rho z$.

An *equivalence relation* on X is a reflexive, symmetric and transitive binary relation.

Definition 1.2.2. A *partially ordered set* or *poset* is a pair $(X, \leq)$ where $\leq$ is a reflexive, antisymmetric, and transitive relation on X.

Definition 1.2.3. Let S be a regular semigroup. For $a, b \in S$, define a partial order as follows

$$a \leq b \text{ iff } a = eb = bf \text{ for some } e, f \in E(S).$$

A partial order is *compatible* if

$$a \leq b \text{ implies } ac \leq bc \text{ and } ca \leq cb \text{ for all } c \in S.$$

This partial order is called the *natural partial order.* See [5].

An ordered semigroup is a semigroup together with a partial order $\leq$ which is compatible.

The partial order in Definition 1.2.3 has several equivalent definitions on a regular semigroup which are taken from [12].

Lemma 1.2.1. *For a regular semigroup $(S, \cdot)$ the following are equivalent:*

(i) $e = eb = bf$ *for some* $e, f \in E(S)$,

(ii) $a = aa'b = ba''a$ *for some* $a', a'' \in V(a) = \{x \in S \mid a = axa, x = xax\}$,

(iii) $a = aa^0b = ba^0a$ *for some* $a^0 \in V(a)$,

(iv) $a'a = a'b$ *and* $aa' = ba'$ *for some* $a' \in V(a)$,

(v) $a = ab^*b = bb^*a,\ a = ab^*a$ *for some* $b^* \in V(b)$,

(vi) $a = axb = bxa,\ a = axa,\ b = bxb$ *for some* $x \in S$,

(vii) $a = eb$ *for some idempotent* $e \in R_a$ *and* $aS \subseteq bS$,

(viii) for every idempotent $f \in R_b$ *there is an idempotent* $e \in R_a$ *with* $e \leq f$ *and* $a = eb$,

(ix) $a = ab'a$ *for some* $b' \in V(b)$, $aS \subseteq bS$ *and* $Sa \subseteq Sb$,

(x) $a = xb = by,\ xa = a$ *for some* $x, y \in S$,

(xi) $a = eb = bx$ *for some* $e \in E(S),\ x \in S$.

Proof. See [12]. □

Now we consider the partial order on Clifford semigroups, we use the definition (vi) above.

Theorem 1.2.1. *Let $S = [Y; G_\alpha, \varphi_{\alpha,\beta}]$ be a Clifford semigroup. Take $a \in G_\alpha, b \in G_\beta$, $\alpha, \beta \in Y$. Then $a \leq b$ if and only if $\alpha \leq \beta$ and $\varphi_{\beta,\alpha}(b) = a$.*

Proof. Sufficiency. Let $\varphi_{\beta,\alpha}(b) = a$ and $\alpha \leq \beta$. Since $a^{-1} \in G_\alpha$ and $aa^{-1}a = a$ and $ba^{-1} = \varphi_{\beta,\alpha}(b)a^{-1} = aa^{-1}$ and $a^{-1}b = a^{-1}\varphi_{\beta,\alpha}(b) = a^{-1}a$ we get $a \leq b$.

Necessity. Let $a \leq b$. Then an element $x \in G_\gamma \subseteq S$ exists for some $\gamma \in Y$ such that $axa = a$, $xa = xb$ and $ax = bx$. So from $axa = a$ we get $\alpha \leq \gamma$ and $ax = bx$ implies $a = axa = bxa$. We have $\alpha \leq \beta$.

From $ax = bx$ it follows that $e_\alpha a e_\gamma = e_\alpha axx^{-1} = e_\alpha bxx^{-1} = e_\alpha b e_\gamma$, thus $a = (e_\alpha \varphi_{\beta,\alpha}(b))\varphi_{\gamma,\alpha}(e_\gamma) = \varphi_{\beta,\alpha}(b)$. □

To illustrate the Theorem 1.2.1 we use the following example.

Example 1.2.1. Consider the Clifford semigroup $S = \mathbb{Z}_{2_\nu} \cup \mathbb{Z}_{2_\alpha} \cup \mathbb{Z}_{3_\gamma} \cup \mathbb{Z}_{4_\beta}$, with a 4-element semilattice $Y = \{\nu < \alpha < \beta, \gamma\}$ and $\mathbb{Z}_n = \{0, 1, \ldots, n-1\}$ the group with addition modulo $n, n \in \{2, 3, 4\}$ such that Hasse diagram is shown below. The defining homomorphisms are according to Theorem 1.2.1, the lines indicate the images of elements under the defining homomorphisms.

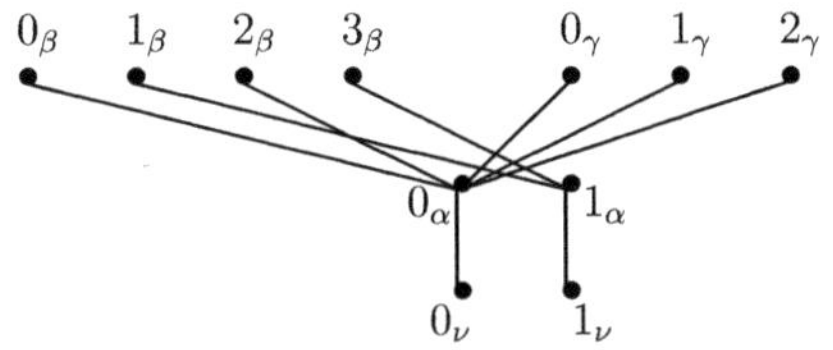

So we have $0_\nu \leq 2_\beta$ since $\varphi_{\beta,\nu}(2_\beta) = 0_\nu$. Thus $0_\nu \leq 0_\xi$ for all $\xi \in Y$, $0_\nu \leq 2_\gamma$, $1_\alpha \leq 1_\beta$ and so on.

Definition 1.2.4. A mapping $f \in End(S)$ is called *order-preserving* homomorphism if $a \leq b$ implies $f(a) \leq f(b)$ with respect to the partial order in Theorem 1.2.1.

We denote by $OEnd(S)$ the monoid of order-preserving endomorphisms of S.

Definition 1.2.5. Let S be a semigroup and $a \in E(S)$. A mapping $c_a \in End(S)$ is defined by $c_a(x) = a$ for all $x \in S$ is called a *constant mapping.*

The following lemma is also true for the case of strong semilattices of left simple semigroups which will come later.

Lemma 1.2.2. *Let $S = [Y; G_\alpha, \varphi_{\alpha,\beta}]$ be a Clifford semigroup and $f \in End(S)$. Then for each $\alpha \in Y$, $f(G_\alpha) \subseteq G_\beta$ for some $\beta \in Y$.*

Proof. Let $x, x^{-1} \in G_\alpha$ be such that $f(x) \in G_\beta$, $f(x^{-1}) \in G_\gamma$ and $f(e_\alpha) = e_\delta$, $\beta, \gamma, \delta \in Y$. Then

$$e_\delta = f(e_\alpha) = f(xx^{-1}) = f(x)f(x^{-1}) \in G_{\beta\gamma}.$$

This implies that $\delta = \beta\gamma \leq \beta$. From $f(x) = f(xe_\alpha) = f(x)f(e_\alpha) = f(x)e_\delta \in G_{\beta\delta}$, we get $\beta = \beta\delta \leq \delta$. This implies $\delta = \beta$.

From $\beta = \delta = \beta\gamma \leq \gamma$ and $f(x^{-1}) = f(x^{-1}e_\alpha) = f(x^{-1})f(e_\alpha) = f(x^{-1})e_\beta \in G_{\gamma\beta}$, and therefore $\gamma = \gamma\beta \leq \beta$. Consequently, $\beta = \gamma = \delta$. □

Definition 1.2.6. From Lemma 1.2.2, for $f \in End(S)$, the mapping $\underline{f} \in End(Y)$ such that $f(G_\alpha) \subseteq G_{\underline{f}(\alpha)}$ is called the *induced index mapping.*

We write the *restriction* $f_\alpha = f|_{G_\alpha}$ of f with the usual meaning. For each $\alpha \in Y$, $f_\alpha \in Hom(G_\alpha, G_{\alpha'})$, we write $f_\alpha(x_\alpha)$ which implies that x_α is considered in G_α, and $f(x_\alpha)$ if f is defined on all of S such that $f(x_\alpha) \in G_{\alpha'}$.

We will show in the following theorem that $OEnd(S) = End(S)$ for a Clifford semigroup S.

Theorem 1.2.2. *Let $S = [Y; G_\alpha, \varphi_{\alpha,\beta}]$ be a Clifford semigroup. Then $OEnd(S) = End(S)$.*

Proof. Take $f \in End(S), a_\alpha \in G_\alpha, b_\beta \in G_\beta$ with $a_\alpha \leq b_\beta$. By Theorem 1.2.1, we get $\varphi_{\beta,\alpha}(b_\beta) = a_\alpha$ and $\alpha \leq \beta$. Suppose that $f(a_\alpha) = x_{\alpha'} \in G_{\alpha'}$ and $f(b_\beta) = y_{\beta'} \in G_{\beta'}$. From $e_\alpha = \varphi_{\beta,\alpha}(e_\beta) = \varphi_{\beta,\alpha}(b_\beta {b_\beta}^{-1}) = \varphi_{\beta,\alpha}(b_\beta)\varphi_{\beta,\alpha}({b_\beta}^{-1}) = a_\alpha {b_\beta}^{-1}$ we get $e_{\alpha'} = f(e_\alpha) = f(a_\alpha {b_\beta}^{-1}) = f(a_\alpha)f({b_\beta}^{-1}) = (x_{\alpha'})(y_{\beta'})^{-1}$, so that $x_{\alpha'} = y_{\beta'}e_{\alpha'}$. We have $\varphi_{\beta',\alpha'}(y_{\beta'}) = x_{\alpha'}$, therefore $f(a_\alpha) = x_{\alpha'} \leq y_{\beta'} = f(b_\beta)$ by Theorem 1.2.1. Therefore $f \in OEnd(S)$ and consequently $End(S) = OEnd(S)$. □

Problem 1.2.1. It would be interesting to investigate orders and order preserving endomorphism of strong semilattices of more general semigroups.

1.3 Regular homomorphisms of groups

In many papers regular endomorphisms of various structures have been studied, for example, the regular endomorphism monoid of groups has been considered in [11], idempotents of endomorphism monoids of groups have been investigated in [17]. In this section, we study homomorphisms in $Hom(G, H)$ which have a "semigroup inverse", which we will introduce. This is a relatively unusual access since $Hom(G, H)$ is not a semigroup (with composition) if $G \not\cong H$. The ideas are based on [11].

Definition 1.3.1. An element f in $Hom(G, H)$ is called *homomorphism regular* if there exists $f' \in Hom(H, G)$ such that $ff'f = f$. The set $Hom(G, H)$ is called *hom-regular* if all its elements are homomorphism regular. An element f' such that $ff'f = f$ and $f' = f'ff'$ is an *inverse* of f.

The set $Hom(\mathbb{Z}_1, \mathbb{Z}_4)$ is regular since there is only the constant map, but the set $Hom(\mathbb{Z}_2, \mathbb{Z}_4)$ is not regular since $f \in Hom(\mathbb{Z}_2, \mathbb{Z}_4)$ with $f(1) = 1$ has no an inverse.

We recall some definitions and notations which are taken from [11].

Definition 1.3.2. Let G be a group. We say that G is a *normal direct sum* of N by K, denoted by $G = N \ltimes K$ if $G = NK$ and $N \cap K = \{e\}$ where e is the identity of G, $N \trianglelefteq G$ (N is a normal subgroup of G), K is a subgroup of G. In this situation we say that K has a *normal complement* in G, and N has a *complement* in G.

If $a \in G = N \ltimes K$, $a = nk$, $n \in N$, $k \in K$, then the map $\pi_K \in End(G)$ defined by $\pi_K(a) = k$ is an idempotent endomorphism of G.

The following results are a generalization of Lemma 1.1 and Theorem 1.2 of [11].

Lemma 1.3.1. *Let G and H be groups and let $f \in Hom(G,H)$ have an inverse f'. Then*

$$Ker(f) = Ker(f'f),\ Im(f) = Im(ff')$$

$$Ker(f') = Ker(ff'),\ Im(f') = Im(f'f).$$

Proof. First $x \in Ker(f)$ implies $e_G = f'(e_H) = f'f(x)$, that is $x \in Ker(f'f)$ i.e., $Ker(f) \subseteq Ker(f'f)$. On the other hand, let $x \in Ker(f'f)$ implies $e_H = f(e_G) = f(f'f(x)) = (ff'f)(x) = f(x)$ that is $x \in Ker(f)$ i.e., $Ker(f'f) \subseteq Ker(f)$. We get $Ker(f) = Ker(f'f)$.

Let $x \in Im(f)$, then $x = f(y)$ for some $y \in G$. We have $x = f(y) = (ff'f)(y) = ff'(f(y)) = ff'(x)$ which $x \in Im(ff')$ i.e., $Im(f) \subseteq Im(ff')$. On the other hand, let $x \in Im(ff')$, then $x = f(f'(y)) \in Im(f)$ for some $y \in H$ and $f'(y) \in G$ i.e., $Im(ff') \subseteq Im(f)$. We get $Im(f) = Im(ff')$.

Let $x \in Ker(f')$, then $e_H = f(e_G) = f(f'(x)) = (ff')(x)$, that is $x \in Ker(ff')$ i.e., $Ker(f') \subseteq Ker(ff')$. On the other hand, $x \in Ker(ff')$ implies $e_G = f'(e_H) = f'(ff'(x)) = (f'ff')(x) = f'(x)$ that is $x \in Ker(f')$ i.e., $Ker(ff') \subseteq Ker(f')$. We get $Ker(f') = Ker(ff')$.

Let $x \in Im(f')$, then $x = f'(y)$ for some $y \in H$. We have $x = f'(y) = (f'ff')(y) = f'f(f'(y)) = f'f(x)$ which $x \in Im(f'f)$ i.e., $Im(f') \subseteq Im(f'f)$. On the other hand, let $x \in Im(f'f)$, then $x = f'f(y) \in Im(f')$ for some $y \in G$ and $f(y) \in H$ i.e., $Im(f'f) \subseteq Im(f')$. We get $Im(f') = Im(f'f)$. □

Theorem 1.3.1. *Let $f \in Hom(G,H)$. Then f has an inverse if and only if $Ker(f)$ has a complement in G and $Im(f)$ has a normal complement in H.*

Proof. Necessity. Let f have an inverse, i.e., there exists $f' \in Hom(H,G)$ such that $ff'f = f$ and $f'ff' = f'$. We now show that $Ker(f)$ has a complement in

G that is $G = Ker(f) \ltimes Im(f')$. Let $x \in Ker(f) \cap Im(f')$, then $f(x) = e_H$ and $x = f'(y)$ for some $y \in H$. It follows that $e_H = f(x) = ff'(y)$ then we have $y \in Ker(ff') = Ker(f')$ by Lemma 1.3.1. Therefore $x = f'(y) = e_G$. Hence $Ker(f) \cap Im(f') = \{e_G\}$.

For all $x \in G$ we have $f(x) = ff'f(x)$ and for $x^{-1} \in G$ we have $e_H = f(e_G) = f(xx^{-1}) = f(x)f(x^{-1}) = f(x)(ff'f)(x^{-1}) = f(xf'f(x^{-1}))$ which means that $xf'f(x^{-1}) \in Ker(f)$. Thus for each $x \in G$ we get $x = xe_G = x(f'f(x^{-1}x)) = (xf'f(x^{-1}))(f'f(x)) \in Ker(f)Im(f')$. Thus $G = Ker(f)Im(f')$. Hence $G = Ker(f) \ltimes Im(f')$. i.e., $Ker(f)$ has a complement in G.

We next show that $Im(f)$ has a normal complement in H, that is $H = Im(f) \ltimes Ker(f')$. Let $x \in Im(f) \cap Ker(f')$, we get $x = f(y)$ and $f'(x) = e_G$ for some $y \in G$. It follows that $e_G = f'(x) = f'(f(y))$ which means that $y \in Ker(f'f) = Ker(f)$ by Lemma 1.3.1. Then $x = f(y) = e_H$, so $Im(f) \cap Ker(f') = \{e_H\}$.

For all $x \in H$ we have $f'(x) = f'ff'(x)$ and for $x^{-1} \in H$ we have $e_G = f'(e_H) = f'(xx^{-1}) = f'(x)f'(x^{-1}) = f'(x)(f'ff')(x^{-1}) = f'(xff'(x^{-1}))$ implies $xff'(x^{-1}) \in Ker(f')$. Thus for each $x \in H$, we get $x = xe_H = x(ff'(e_H)) = x(ff'(x^{-1}x)) = (xff'(x^{-1}))(ff'(x)) \in Ker(f')Im(f)$. Thus $H = Ker(f')Im(f)$. Hence $H = Ker(f') \ltimes Im(f)$. i.e., $Im(f)$ has a normal complement in H.

Sufficiency. Let $G = Ker(f) \ltimes K$ and $H = Im(f) \ltimes N$ where K is a subgroup in G and N is a normal subgroup in H. We note that $K \cong G/Ker(f) \cong Im(f)$. So we can define $\phi : K \to Im(f)$ by $\phi(k) = f(k)$ for every $k \in K$. Define $f' : H \to G$ by $f'(h) = (\phi^{-1}\pi_{Im(f)})(h)$ where $\pi_{Im(f)} : H \to Im(f)$ is the projection onto $Im(f)$. For each $x \in G = Ker(f) \ltimes K$ we have $x = yz$ for some $y \in Ker(f)$ and $z \in K$, so $ff'f(x) = ff'(f(yz)) = ff'(f(z)) = f\phi^{-1}\pi_{Im(f)}(f(z)) = f\phi^{-1}f(z) = f(z) = f(x)$. We note that $Im(f) = Im(\phi)$ and $K = Im(\phi^{-1})$. We also have

$$f'ff' = f'\phi\pi_K\phi^{-1}\pi_{Im(f)} = f'\phi^{-1}\phi = f',$$

where $\pi_K : G \to K$ is the projection onto K. Hence f' is an inverse of f. □

In the case that G is a commutative group, the direct product was mentioned as a direct sum (see [9]).

Definition 1.3.3. A subgroup A of a group G is called a *direct sum* of G if there is a subgroup B of G such that $A \cap B = \{e\}$ and $A + B = G$, where e is the identity in G. We write $G = A \oplus B$ as G is a a direct sum of A and B.

In [9], they found that the endomorphism ring of an abelian group G is regular if and only if images and kernels of all endomorphisms of G are direct sums of G and the regularity of endomorphisms of modules can be found in [11].

The following example shows that the set $Hom(\mathbb{Z}_6, \mathbb{Z}_4)$ is not regular and the set $Hom(\mathbb{Z}_3, \mathbb{Z}_6)$ is regular.

Example 1.3.1. Consider the set of homomorphisms of $Hom(\mathbb{Z}_6, \mathbb{Z}_4)$. Take $f \in Hom(\mathbb{Z}_6, \mathbb{Z}_4)$ as follows

$x \in \mathbb{Z}_6$	0	1	2	3	4	5
$f(x) \in \mathbb{Z}_4$	0	2	0	2	0	2.

It can be seen that $Im(f) = \{0, 2\} \subseteq \mathbb{Z}_4$, which is not a direct sum of $\mathbb{Z}_4$ while $Ker(f) = \{0, 2, 4\} \subseteq \mathbb{Z}_6$ is a direct sum of $\mathbb{Z}_6$. Then we have that f is not regular by Theorem 1.3.1.

Now, take $g \in Hom(\mathbb{Z}_3, \mathbb{Z}_6)$ as follows

$x \in \mathbb{Z}_3$	0	1	2
$g(x) \in \mathbb{Z}_6$	0	4	2.

It can be seen that $Im(g) = \{0, 2, 4\} \subseteq \mathbb{Z}_6$ such that $\mathbb{Z}_6 = Im(g) \times \{0, 3\}$, and $Ker(g) = \{0\} \subseteq \mathbb{Z}_3$ such that $\mathbb{Z}_3 = Ker(g) \times \mathbb{Z}_3$. Then g is regular by Theorem 1.3.1.

We note that in the commutative case the direct product $\ltimes$ is written as $\times$.

The next result has been proved in [11].

Corollary 1.3.1. *A group G, $End(G)$ is regular if and only if every kernel of an endomorphism has a complement and every image of an endomorphism has a normal complement.*

Example 1.3.2. The monoid $End(\mathbb{Z}_6)$ is regular since all subgroups of $\mathbb{Z}_6$ are $\mathbb{Z}_3$ and $\mathbb{Z}_2$ such that each has a complement subgroup. Let Q be the quaternion group with

$$\mathrm{Q} = < a, b \mid a^4 = e, b^2 = a^2, ba = a^3b > .$$

The monoid $End(\mathrm{Q})$ is not regular. To see this, take $f \in End(\mathrm{Q})$ as follows

$$f = \begin{pmatrix} e & a & a^2 & a^3 & b & ab & a^2b & a^3b \\ e & e & e & e & a^2 & a^2 & a^2 & a^2 \end{pmatrix}.$$

We have $Im(f) = \{e, a^2\}$ has no complement subgroup in Q. This implies that f is not regular by Corollary 1.3.1.

Let $\mathbb{Z}_n$ be the cyclic group of order n. Then the endomorphism ring $(End(\mathbb{Z}_n), \circ, +) \cong (\mathbb{Z}_n, \cdot)$. (See [9]).

We collect groups whose endomorphism monoids are regular, idempotent-closed, and completely regular. For the finite cyclic group $\mathbb{Z}_n$ we know from [1] that the endomorphism monoid of a finite cyclic group is regular if and only if the order of the group is square-free. For the other groups, for example the symmetric group S_3, the quaternion group Q we have calculated ourselves.

$End(G)$	regular	idem-closed	completely regular
$\mathbb{Z}_p$	✓	✓	✓
$\mathbb{Z}_4$	×	✓	×
$\mathbb{Z}_6$	✓	✓	✓
$\mathbb{Z}_8$	×	✓	×
$\mathbb{Z}_2 \times \mathbb{Z}_2$	✓	×	×
S_3	✓	✓	✓
Q	×	✓	×

Chapter 2

Endomorphisms of semilattices

2.1 Finite semilattices with regular endomorphisms

In this section we provided all of definitions, terminology and property for finite semilattices whose endomorphism monoids are regular which are investigated in [2]. Before we state the main result of [2], some definitions and notations are needed (see also [2]).

Definition 2.1.1. Elements a, b of a semilattice S are *comparable* if $a \wedge b \in \{a, b\}$, if a and b are not comparable, we write $a \| b$. A $\wedge$-*reducible* element is one that can be expressed as $a \wedge b$ where $a \| b$. If S is a lattice, then a $\vee$-*reducible* element is one that can be written as $a \vee b$ for some $a \| b$. A subset C of S for which all $a, b \in C$ are comparable is called a *chain*. An *antichain* is a subset A of S such that $a \| b$ for all distinct $a, b \in A$.

Definition 2.1.2. For an element a of a semilattice Y, the *principal ideal* generated by a is the set $(a] = \{x \in Y \mid x \leq a\}$, and the *principal filter* generated by a is the set $[a) = \{x \in Y \mid x \geq a\}$. If $(a]$ is a chain for all $a \in Y$, then Y is said to be a *tree*. A tree is said to be *binary* if for each $\wedge$-reducible a there are precisely two elements that cover a. An element a is said to be *cover* an element b, denoted $a \succ b$ if $a > b$ and there is no c satisfying $a > c > b$.

The following figure is an example of a binary tree that 0 is a $\wedge$-reducible element.

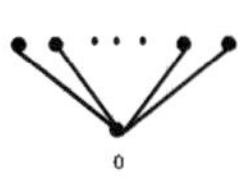

Definition 2.1.3. A semilattice Y is said to satisfying the *strong meet property* if $a_0 \wedge a_1 = b_0 \wedge b_1$ whenever a_0, a_1, b_0, b_1 are elements of Y such that $a_0 \| a_1$ and $b_i \in [a_i) \backslash [a_{1-i})$ for $i = 0, 1$.

Note that if Y is a tree, it is equivalent to assert that $a_0 \wedge a_1 = b_0 \wedge b_1$ whenever $a_0 \| a_1$ and $b_i \in [a_i)$, $i = 0, 1$.

Lemma 2.1.1. *Every tree satisfies the strong meet property.*

Now definitions of the classes $\mathbf{B}$ and $\mathbf{B}^d$ are provided.

Definition 2.1.4. A *capped binary tree* is the lattice obtained by adjoining a unit, ie.e., a greatest element to a binary tree. The *vertical sum* of bounded lattices L_0 and L_1 is defined (only up to isomorphism) by first replacing each L_i by an isomorphic copy L_i' such that the unit of L_0' is the zero, i.e., the smallest element of L_1' and is the only element of $L_0' \cap L_1'$. A partial order is then defined on $L_0' \cup L_1'$ by retaining the ordering within each lattice and stipulating that $x \leq y$ whenever $x \in L_0'$ and L_1'. The resulting lattice is denoted $L_0 +_V L_1$.

In practice, the distinction between L_i and L_i' will be suppressed and L_i will be regarded as a sublattice of $L_0 +_V L_1$.

Given bounded lattices $L_i, i < n$ where $n > 1$, the vertical sum $\sum_V(L_i, i < n)$ is defined to be $\cdots((L_0 +_V L_1) +_V L_2) +_V \cdots) +_V L_{n-1}$.

The following figure is an example of elements of the class $\mathbf{B}$.

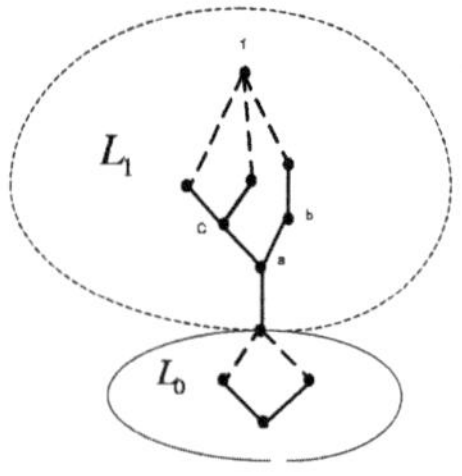

$L_0 +_V L_1$

Definition 2.1.5. Let $\mathbf{B}$ denote the class of all vertical sums of finite capped binary trees, and let $\mathbf{B}^d$ denote the class of all lattices L such that the dual of L lies in $\mathbf{B}$.

Definition 2.1.6. Let Y be a finite lattice. A subsemilattice of $(Y; \wedge)$ is said to be a $\wedge$-*subsemilattice* of Y. A bounded $\wedge$- subsemilattice T of Y is said to be *smooth* if T

does not contain elements a, b, c satisfying $c \| a \vee b$ and $c < a \vee_T b$, where $\vee_T$ denotes join with respect to T.

Lemma 2.1.2. *Let L_i, $i < n$, where $n > 1$, be finite lattices, and let $Y = \sum_V(L_i, i < n)$. If each L_i satisfies the strong meet property, then so does Y.*

Proof. See [2] Lemma 4.1. □

Moreover, **R** denote the intersection of all rectilinearly closed classes that contain the one-element and two-element chains.

The following figure is an example of elements of the class $\mathbf{B}^d$. We observe that the figure turns the element of the class **B** down.

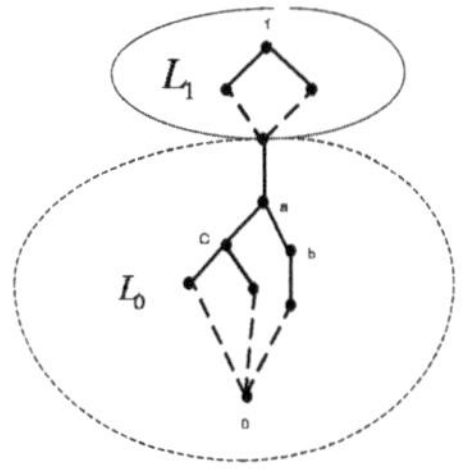

$L_0 +_V L_1$

Now definitions for the class **R** are provided.

Definition 2.1.7. Given bounded lattices L_i, $i < n$, where $n > 1$, their *horizontal sum* is defined (only up to isomorphism) as follows. First replace each L_i by an isomorphic copy L'_i such that $L'_i \cap L'_j = \emptyset$ whenever $i \neq j$, and choose $0, 1$ to be any objects not elements of $\cup(L'_i, i < n)$. A partial order is then defined on $\cup(L'_i, i < n) \cup \{0, 1\}$ by retaining the ordering within each lattice and defining $0 < x < 1$ for all $x \in \cup(L'_i, i < n)$. The resulting lattice is denoted by $\sum_H(L_i, i < n)$.

In practice, the distinction between L_i and L'_i will be suppressed, that is, the L_i will be presumed pairwise disjoint.

Definition 2.1.8. A class **K** of finite lattices is said to be *rectilinearly closed* if, for all $n > 1$, $\sum_V(L_i, i < n)$ and $\sum_H(L_i, i < n)$ both belong to **K** whenever $L_i \in \mathbf{K}$, $i < n$.

Let **R** denote the intersection of all rectilinearly closed classes that contain the one-element and two-element chains.

Definition 2.1.9. An antichain A in a lattice Y is said to be *self-disjoint* if $a_0 \wedge a_1 = b_0 \wedge b_1$ whenever a_0, a_1, b_0, b_1 are elements of A with $a_0 \neq a_1$ and $b_0 \neq b_1$. We say that Y satisfies *strong antichain property* if every antichain in Y is self-disjoint or contains distinct elements a, b, c such that $a \wedge (b \vee c) \leq b$.

Proposition 2.1.1. *For every $Y \in \mathbf{R}$ the following hold.*

1) Y satisfies the strong meet property.

2) Y satisfies the strong antichain property.

3) Every smooth $\wedge$-subsemilattice of Y is a member of **R**.

Proof. See [2] Lemma 5.1. □

The left figure is an element of the class **R**, but the right figure is not because for an antichain subset $A = \{a, b, d\}$ of Y, $a \wedge d = 0$ but $a \wedge b = c$. This is a contradiction with the property of **R** in Proposition 2.1.1.

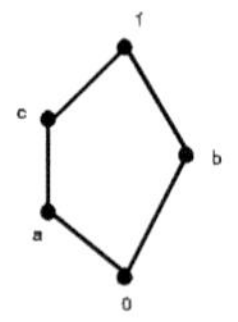

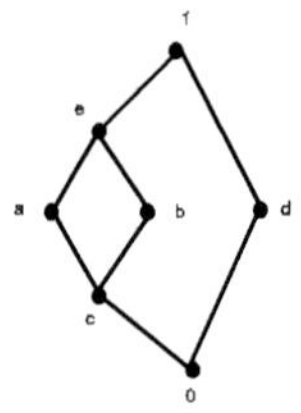

The next theorem is the main results of [2].

Theorem 2.1.1. *For a finite semilattice Y, End(Y) is regular if and only if one of the following holds*

1) Y is a binary tree,

2) Y is a tree with one $\wedge$-reducible element, or

3) Y is a bounded lattice, $Y \in \mathbf{B} \cup \mathbf{B}^d \cup \mathbf{R}$.

2.2 Properties of endomorphisms of semilattices and sets

In this section we investigate the properties from Definition 1.1.2, 1.1.6 and Definition 1.1.7, namely, idempotent-closed, orthodox, left inverse, completely regular and idempotent, of endomorphism monoids of finite semilattices and of sets.

Lemma 2.2.1. *Let Y be a finite semilattice and let $s \in End(Y)$, $\alpha, \beta, \gamma \in Y$.*

1) If $\alpha < \beta < \gamma$ and $s(\alpha) = s(\gamma) = \delta$ for some $\delta \in Y$, then $s(\beta) = \delta$.

2) If $\alpha = \beta\gamma$ where $\beta \| \gamma$ and $s(\beta) = s(\gamma) = \delta$ for some $\delta \in Y$, then $s(\alpha) = \delta$.

Proof. 1) $s(\beta) = s(\beta\gamma) = s(\beta)s(\gamma) = s(\beta)\delta \leq \delta$ and

$$\delta = s(\alpha) = s(\alpha\beta) = s(\alpha)s(\beta) = \delta s(\beta) \leq s(\beta).$$

This implies $s(\beta) = \delta$.

2) $s(\alpha) = s(\beta\gamma) = s(\beta)s(\gamma) = \delta\delta = \delta$. □

By $Y = Y_{0,n}$ denote the semilattice with minimum 0 and the graph structure of the complete bipartite graph $K_{1,n}$. See the figure of $K_{1,3}$ as follows.

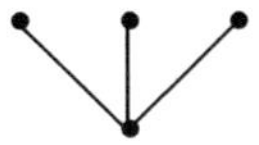

Lemma 2.2.2. *Take $s \in End(Y_{0,n})$ not constant. Then s is idempotent if and only if $s(\alpha) \neq \alpha$ implies $s(\alpha) = 0$ for $\alpha \in Y_{0,n}$.*

Proof. Necessity. Suppose that $s(\alpha) = \beta$ for some $\beta \neq \alpha$. Then $s(\beta) = s(s(\alpha)) = s(\alpha) = \beta$. Thus $0 = s(0) = s(\alpha\beta) = s(\alpha)s(\beta) = \beta\beta = \beta$ and therefore $s(\alpha) = 0$.

Sufficiency. For each $\alpha \in Y_{0,n}$, if $s(\alpha) = \alpha$, then $s(s(\alpha)) = s(\alpha)$.
If $s(\alpha) \neq \alpha$, then $s(\alpha) = 0$ by hypothesis, so that $ss(\alpha) = s(0) = 0$ and $s(\alpha) = 0$.
Therefore s is idempotent. □

Lemma 2.2.3. *Let Y be a finite semilattice. Then $End(Y)$ is idempotent-closed if and only if $Y = Y_{0,n}$.*

Proof. Necessity. Suppose that Y contains a chain $\{1,2,3\}$. Take two idempotents $s, t \in End(Y)$ such that $s(1) = s(2) = 2, s(3) = 3$ and $t(1) = 1, t(2) = t(3) = 3$ and then $(st)^2(1) = 3$ but $(st)(1) = 2$. Thus st is not an idempotent. This implies that Y does not contain a chain. Since Y is a semilattice, we have $1 \wedge 2 = 3$, so that $Y = Y_{o,n}$.

Sufficiency. Take two idempotents $s, t \in End(Y_{0,n})$. We use Lemma 2.2.2, and consider two cases,

if s or t is constant, then (st) is constant and of course, it is idempotent,

if s and t are not constant, then for each $\alpha \in Y_{0,n}$, $st(\alpha) = \alpha$ if $s(\alpha) = \alpha$ and $t(\alpha) = \alpha$. Further, for $s(\alpha) \neq \alpha$ or $t(\alpha) \neq \alpha$, we have $stst(\alpha) = 0 = s(t(\alpha))$,

Thus st is idempotent, and therefore $End(Y_{0,n})$ is idempotent-closed. □

We now consider the monoid $End(Y)$ for a finite semilattice Y.

Proposition 2.2.1. *Let Y be a finite semilattice. Then the monoid $End(Y)$ is*

1) *regular if and only if Y is a binary tree or a tree with one $\wedge$-reducible or $Y \in \mathbf{B} \cup \mathbf{B}^d \cup \mathbf{R}$ (see Theorem 2.1.13).*

2) *completely regular*
3) *idempotent*

} *if and only if $|Y| \leq 2$.*

4) *idempotent-closed*
5) *orthodox*
6) *left inverse*

} *if and only if $Y = Y_{0,n}$.*

7) *right inverse*
8) *inverse*
9) *a group*
10) *commutative*

} *if and only if $|Y| = 1$.*

Proof. Necessities.

1) is taken from [2].

4) and 5) follow from Lemma 2.2.3.

The statements 7), 8), 9) and 10) are trivial.

We verify 2) and 3). Suppose that Y contains a chain $\{1,2,3\}$ or a semilattice $1 \wedge 2 = 3$. Take $s \in End(Y)$ such that $s(1) = 2, s(2) = s(3) = 3$. Then any $t \in End(Y)$ such that $sts = s$, t must fulfill $t(2) = 1$ and then $ts(1) = 1$ but $1 \notin Im(st)$, so s is not completely regular, and therefore $End(Y)$ is not completely regular. It can be seen that s is not idempotent. This follows that $End(Y)$ is not idempotent. Hence $|Y| \leq 2$.

6) Suppose that Y contains a chain $\{1, 2, 3\}$. Take two idempotents $s, t \in End(Y)$ such that $s(1) = 1, s(2) = s(3) = 3$ and $t(1) = t(2) = 2, t(3) = 3$. Then $tst(1) = tst(2) = tst(3) = 3$ but $ts(1) = 2$. Thus $tst \neq ts$, and therefore $End(Y)$ is not left inverse. Since Y is a semilattice, we have $1 \wedge 2 = 3$. Hence $Y = Y_{0,n}$.

Sufficiency. If $|Y| = 1$, then everything is obvious.

2) and 3) Take $|Y| = 2$. Then $End(Y)$ consists of two constant maps and the identity map.

6) Take $Y = Y_{0,n}$. We now show that $End(Y_{0,n})$ is left inverse. Take two idempotents $s, t \in End(Y_{0,n})$. By using Lemma 2.2.2, we have.

If s or t is constant, then we have $sts = st$.

If s and t are not constant. Then

$$sts(\alpha) = \begin{cases} \alpha & \text{if } s(\alpha) = \alpha \text{ and } t(\alpha) = \alpha, \\ 0 & \text{otherwise.} \end{cases}$$

Thus $sts = st$, and therefore $End(Y_{0,n})$ is left inverse. □

As a consequence of Proposition 2.2.1 we have:

Corollary 2.2.1. *Let Y be a non-trivial finite chain. Then $End(Y)$ is*

1) *always regular.*

$$\left.\begin{array}{l} \text{2) completely regular} \\ \text{3) idempotent} \\ \text{4) idempotent-closed} \\ \text{5) orthodox} \\ \text{6) left inverse} \end{array}\right\} \text{ if and only if } |Y| \leq 2$$

$$\left.\begin{array}{l} \text{7) right inverse} \\ \text{8) inverse} \\ \text{9) a group} \\ \text{10) commutative} \end{array}\right\} \text{ if and only if } |Y| = 1.$$

Corollary 2.2.2. *Any finite semilattice Y such that $End(Y)$ satisfies any one of the properties of Proposition 2.2.1, does not contain a three-element chain.*

As a consequence we get most of the next corollary which has also been formulated in [7].

Corollary 2.2.3. *Let A be a set. Consider the monoid $T(A)$ of all mappings of A into itself. $T(A)$ is*

1) *always regular.*

2) *completely regular*
3) *idempotent-closed*
4) *orthodox*
5) *left inverse*

} *if and only if* $|A| \leq 2$

6) *right inverse*
7) *inverse*
8) *a group*
9) *commutative*
10) *idempotent*

} *if and only if* $|A| = 1$.

Chapter 3

Endomorphisms of strong semilattices of left simple semigroups

In this chapter we consider the strong semilattices of left simple semigroups in which the defining homomorphisms are constant or bijective whose endomorphism monoids are regular, idempotent-closed, orthodox, left inverse, completely regular and idempotent.

We note that the semilattice Y which is considering, is non-trivial, i.e., $|Y| > 1$.

Let $S = [Y; S_\alpha, e_\alpha, \varphi_{\alpha,\beta}]$ be a non-trivial strong semilattice of semigroups S_α with a fixed idempotent e_α for $\alpha \in Y$ and defining homomorphisms $\varphi_{\alpha,\beta}$ for $\beta \leq \alpha$.

We collect the results of this chapter as a table in the Overview.

3.1 Homomorphisms of a non-trivial strong semilattice of semigroups

Definition 3.1.1. A semigroup S is called *left simple* if $S = Sa$ for all $a \in S$. An analogy, S is called *right simple* if $S = aS$ for all $a \in S$.

We denote by

$$E_\varphi(S) = \{e_\beta \mid \varphi_{\alpha,\beta}(e_\alpha) = e_\beta \text{ for some idempotent } e_\alpha \in S_\alpha,\ \beta \leq \alpha \in Y\}.$$

In general if $f \in End(S)$, then $\underline{f}$ may not be a mapping. To see this, we show the following example. This example is suggested by Professor Norman R. Reilly.

Example 3.1.1. Consider the semilattice $Y = \{0, \alpha, \beta, \gamma, \sigma\}$ as shown below and let $S = [Y; S_\alpha, e_\alpha, \varphi_{\alpha,\beta}]$ be a non-trivial strong semilattice of idempotent semigroups such that $S_\alpha = \{a_\alpha, b_\alpha, c_\alpha\}$ with $a_\alpha b_\alpha = b_\alpha a_\alpha = c_\alpha$, the remaining semigroups consist of one idempotent. Take $f \in End(S)$ as follows.

$$f = \begin{pmatrix} e_0 & a_\alpha & b_\alpha & c_\alpha & e_\beta & e_\gamma & e_\sigma \\ e_0 & e_\gamma & e_\delta & e_\beta & e_0 & e_0 & e_0 \end{pmatrix}.$$

It can be seen that $\underline{f} \notin End(Y)$.

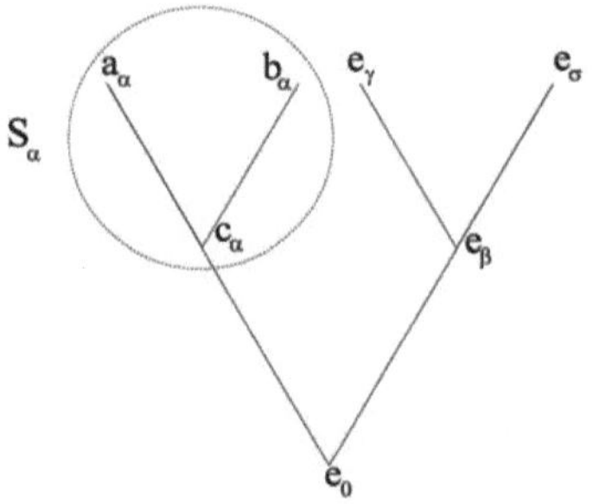

Lemma 3.1.1. *Let S be a simple semigroup with idempotent e. Then $x = xe$ for all $x \in S$.*

Proof. Take any $x \in S$. Since S is a left simple semigroup, we have $S = Se$ and $x = ye$ for some $y \in S$. Then

$$xe = yee = ye = x.$$

This implies that $x = xe$ for all $x, e \in S$. □

The next lemma is a generalization of Lemma 1.3 of [20].

Lemma 3.1.2. *Let $S = [Y; S_\alpha, e_\alpha, \varphi_{\alpha,\beta}]$ be a non-trivial strong semilattice of semigroups. Let $s \in End(Y)$ and let $\{f_\alpha \in Hom(S_\alpha, S_{s(\alpha)}) \mid \alpha \in Y\}$ be a family of semigroup homomorphisms which satisfies*

$$f_\beta \varphi_{\alpha,\beta} = \varphi_{s(\alpha),s(\beta)} f_\alpha$$

for all $\alpha, \beta \in Y$. Then $f : S \to S$ defined by $f(x_\alpha) := f_\alpha(x_\alpha)$ for every $x_\alpha \in S_\alpha$, is an endomorphism on S.

Proof. It can be seen that f is well-defined. We verify now that f is a homomorphism. Take $x_\alpha, y_\beta \in S,\ \alpha, \beta \in Y$. Then

$$\begin{aligned} f(x_\alpha y_\beta) &= f(\varphi_{\alpha,\alpha\beta}(x_\alpha)\varphi_{\beta,\alpha\beta}(y_\beta)) \\ &= f_{\alpha\beta}(\varphi_{\alpha,\alpha\beta}(x_\alpha))f_{\alpha\beta}(\varphi_{\beta,\alpha\beta}(y_\beta)) \\ &= \varphi_{s(\alpha),s(\alpha\beta)}f_\alpha(x_\alpha)\varphi_{s(\beta),s(\alpha\beta)}f_\beta(y_\beta) \\ &= f_\alpha(x_\alpha)f_\beta(y_\beta) \\ &= f(x_\alpha)f(y_\beta). \end{aligned}$$

Then $f \in End(S)$. □

Lemma 3.1.3. *Let $S = [Y; S_\alpha, e_\alpha, \varphi_{\alpha,\beta}]$ and $T = [Z; T_\alpha, e_\alpha, \psi_{\alpha,\beta}]$ be two strong semilattice of left simple semigroups. Let $f : S \to T$ be a homomorphism. Then for $\alpha \in Y$, $f(S_\alpha) \subseteq T_\beta$ for some $\beta \in Y$. That is $\underline{f} \in Hom(Y, Z)$ and $f(e_\alpha) \in E(T_\beta)$.*

Proof. Let $x_\alpha, y_\alpha \in S_\alpha$. Suppose that $f(x_\alpha) \in T_\beta$ and $f(y_\alpha) \in T_\gamma$ for some $\beta, \gamma \in Z$. Since S_α is a left simple semigroup, we have $x_\alpha \in S_\alpha = S_\alpha y_\alpha$ and $y_\alpha \in S_\alpha = S_\alpha x_\alpha$, so that $x_\alpha = a_\alpha y_\alpha$ and $y_\alpha = b_\alpha x_\alpha$ for some $a_\alpha, b_\alpha \in S_\alpha$.

Assume that $f(a_\alpha) \in T_\delta$ and $f(b_\alpha) \in T_\zeta$ for some $\delta, \zeta \in Z$. Then

$$f(a_\alpha y_\alpha) = f(a_\alpha) \in T_\beta$$

and

$$f(a_\alpha)f(y_\alpha) \in T_{\delta\gamma}.$$

This implies that $\beta = \delta\gamma \leq \gamma$.

Now we consider

$$f(b_\alpha x_\alpha) = f(y_\alpha) \in T_\gamma$$

and

$$f(b_\alpha)f(x_\alpha) \in T_{\zeta\beta}.$$

This implies that $\gamma = \zeta\beta \leq \beta$, and therefore $\beta = \gamma$. Hence for each $\alpha \in Y$, $f(S_\alpha) \subseteq T_\beta$ for some $\beta \in Z$. □

Corollary 3.1.1. *Let $S = [Y; S_\alpha, e_\alpha, \varphi_{\alpha,\beta}]$ be a non-trivial strong semilattice of left simple semigroups. Let $f : S \to S$ be an endomorphism of S. Then for $\alpha \in Y$, $f(S_\alpha) \subseteq S_\beta$ for some $\beta \in Y$. That is $\underline{f} \in End(Y)$ and $f(e_\alpha) \in E(S_\beta)$.*

Example 3.1.2. Let $S = S_{3_\nu} \cup \mathbb{Z}_{3_\alpha}$, $\varphi_{\alpha,\nu} = c_{(1)_\nu}$ be a strong semilattice of groups. The diagram is shown below.

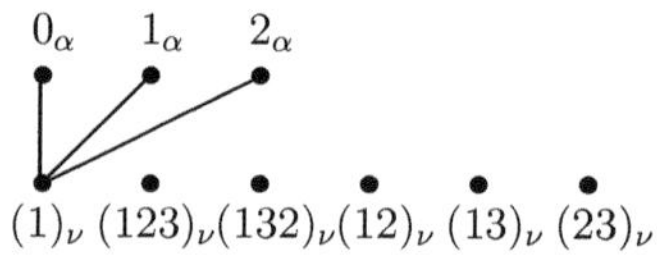

From Lemma 3.1.2 we can construct all endomorphisms of S. On the other hand, if $f \in End(S)$ then $\underline{f} \in End(Y)$ and $f(G_\alpha) \subseteq G_\beta$ for some $\beta \in Y$ by Corollary 3.1.1. All endomorphisms of S are shown below.

	$(1)_\nu$	$(123)_\nu$	$(132)_\nu$	$(12)_\nu$	$(13)_\nu$	$(23)_\nu$	0_α	1_α	2_α	
f_1	$(1)_\nu$	$(1)_\nu$	$(1)_\nu$	$(1)_\nu$	$(1)_\nu$	$(1)_\nu$	$(1)_\nu$	$(1)_\nu$	$(1)_\nu$	$\underline{f}(\alpha) = \underline{f}(\nu) = \nu$
f_2	$(1)_\nu$	$(1)_\nu$	$(1)_\nu$	$(12)_\nu$	$(12)_\nu$	$(12)_\nu$	$(1)_\nu$	$(1)_\nu$	$(1)_\nu$	〃
f_3	$(1)_\nu$	$(1)_\nu$	$(1)_\nu$	$(13)_\nu$	$(13)_\nu$	$(13)_\nu$	$(1)_\nu$	$(1)_\nu$	$(1)_\nu$	〃
f_4	$(1)_\nu$	$(1)_\nu$	$(1)_\nu$	$(23)_\nu$	$(23)_\nu$	$(23)_\nu$	$(1)_\nu$	$(1)_\nu$	$(1)_\nu$	〃
f_5	$(1)_\nu$	$(123)_\nu$	$(132)_\nu$	$(12)_\nu$	$(13)_\nu$	$(23)_\nu$	$(1)_\nu$	$(1)_\nu$	$(1)_\nu$	〃
f_6	$(1)_\nu$	$(132)_\nu$	$(123)_\nu$	$(12)_\nu$	$(23)_\nu$	$(13)_\nu$	$(1)_\nu$	$(1)_\nu$	$(1)_\nu$	〃
f_7	0_α	0_α	0_α	0_α	0_α	0_α	0_α	0_α	0_α	$\underline{f}(\alpha) = \underline{f}(\nu) = \alpha$
f_8	$(1)_\nu$	$(1)_\nu$	$(1)_\nu$	$(1)_\nu$	$(1)_\nu$	$(1)_\nu$	0_α	0_α	0_α	$\underline{f}(\alpha) = \alpha, \underline{f}(\nu) = \nu$
f_9	$(1)_\nu$	$(1)_\nu$	$(1)_\nu$	$(1)_\nu$	$(1)_\nu$	$(1)_\nu$	0_α	1_α	2_α	〃
f_{10}	$(1)_\nu$	$(1)_\nu$	$(1)_\nu$	$(1)_\nu$	$(1)_\nu$	$(1)_\nu$	0_α	2_α	1_α	〃
f_{11}	$(1)_\nu$	$(123)_\nu$	$(132)_\nu$	$(12)_\nu$	$(13)_\nu$	$(23)_\nu$	0_α	0_α	0_α	〃
f_{12}	$(1)_\nu$	$(123)_\nu$	$(132)_\nu$	$(12)_\nu$	$(13)_\nu$	$(23)_\nu$	0_α	1_α	2_α	〃
f_{13}	$(1)_\nu$	$(123)_\nu$	$(132)_\nu$	$(12)_\nu$	$(13)_\nu$	$(23)_\nu$	0_α	2_α	1_α	〃
f_{14}	$(1)_\nu$	$(1)_\nu$	$(1)_\nu$	$(12)_\nu$	$(12)_\nu$	$(12)_\nu$	0_α	0_α	0_α	〃
f_{15}	$(1)_\nu$	$(1)_\nu$	$(1)_\nu$	$(12)_\nu$	$(12)_\nu$	$(12)_\nu$	0_α	1_α	2_α	〃
f_{16}	$(1)_\nu$	$(1)_\nu$	$(1)_\nu$	$(12)_\nu$	$(12)_\nu$	$(12)_\nu$	0_α	2_α	1_α	〃
f_{17}	$(1)_\nu$	$(1)_\nu$	$(1)_\nu$	$(13)_\nu$	$(13)_\nu$	$(13)_\nu$	0_α	0_α	0_α	〃
f_{18}	$(1)_\nu$	$(1)_\nu$	$(1)_\nu$	$(13)_\nu$	$(13)_\nu$	$(13)_\nu$	0_α	1_α	2_α	〃
f_{19}	$(1)_\nu$	$(1)_\nu$	$(1)_\nu$	$(13)_\nu$	$(13)_\nu$	$(13)_\nu$	0_α	2_α	1_α	〃
f_{20}	$(1)_\nu$	$(1)_\nu$	$(1)_\nu$	$(23)_\nu$	$(23)_\nu$	$(23)_\nu$	0_α	0_α	0_α	〃
f_{21}	$(1)_\nu$	$(1)_\nu$	$(1)_\nu$	$(23)_\nu$	$(23)_\nu$	$(23)_\nu$	0_α	1_α	2_α	〃
f_{22}	$(1)_\nu$	$(1)_\nu$	$(1)_\nu$	$(23)_\nu$	$(23)_\nu$	$(23)_\nu$	0_α	2_α	1_α	〃
f_{23}	$(1)_\nu$	$(132)_\nu$	$(123)_\nu$	$(12)_\nu$	$(23)_\nu$	$(13)_\nu$	0_α	0_α	0_α	〃
f_{24}	$(1)_\nu$	$(132)_\nu$	$(123)_\nu$	$(12)_\nu$	$(23)_\nu$	$(13)_\nu$	0_α	1_α	2_α	〃
f_{25}	$(1)_\nu$	$(132)_\nu$	$(123)_\nu$	$(12)_\nu$	$(23)_\nu$	$(13)_\nu$	0_α	2_α	1_α	〃

From now on we use the notation $S = [Y; S_\alpha, e_\alpha, c_{\alpha,e_\beta}]$ instead of a non-trivial strong semilattice of semigroups with constant defining homomorphisms $\varphi_{\alpha,\beta} = c_{\alpha,e_\beta}$.

An endomorphism f of a non-trivial strong semilattice of left simple semigroups with constant defining homomorphisms $\varphi_{\alpha,\beta}$ always satisfies the following equations,

$$f_\beta \varphi_{\alpha,\beta} = \varphi_{\underline{f}(\alpha),\underline{f}(\beta)} f_\alpha$$

for all $\alpha, \beta \in Y$, which is shown in the next lemma.

Lemma 3.1.4. *Let $S = [Y; S_\alpha, e_\alpha, c_{\alpha,e_\beta}]$ be a non-trivial strong semilattice of left simple semigroups. Let $f \in End(S)$. Then $\underline{f} \in End(Y)$ and the set $\{f_\alpha \in Hom(S_\alpha, S_{\underline{f}(\alpha)}) \mid \alpha \in Y\}$ satisfies*

$$f_\beta \varphi_{\alpha,\beta} = \varphi_{\underline{f}(\alpha),\underline{f}(\beta)} f_\alpha$$

for all $\alpha, \beta \in Y$.

Proof. Since $f \in End(S)$, we have $\underline{f} \in End(Y)$ by Corollary 3.1.1 and for $\beta < \alpha \in Y$, take $x_\alpha \in S_\alpha$ and idempotents $e_\beta, e'_\beta \in S_\beta$, suppose that $f(e_\beta) = e'_{\underline{f}(\beta)}$. Since $S_{\underline{f}(\beta)} = S_{\underline{f}(\beta)} e'_{\underline{f}(\beta)}$, we have $e_{\underline{f}(\beta)} = y_{\underline{f}(\beta)} e'_{\underline{f}(\beta)}$ for some $y_{\underline{f}(\beta)} \in S_{\underline{f}(\beta)}$ and

$$e_{\underline{f}(\beta)} e'_{\underline{f}(\beta)} = y_{\underline{f}(\beta)} e'_{\underline{f}(\beta)} e'_{\underline{f}(\beta)} = y_{\underline{f}(\beta)} e'_{\underline{f}(\beta)} = e_{\underline{f}(\beta)}.$$

Thus

$$f(x_\alpha e_\beta) = f(\varphi_{\alpha,\beta}(x_\alpha) e_\beta) = f(e_\beta) = e'_{\underline{f}(\beta)}$$

and

$$f(x_\alpha) f(e_\beta) = \varphi_{\underline{f}(\alpha),\underline{f}(\beta)}(f(x_\alpha)) e'_{\underline{f}(\beta)} = e_{\underline{f}(\beta)} e'_{\underline{f}(\beta)} = e_{\underline{f}(\beta)}.$$

This implies that $f(e_\beta) = e'_{\underline{f}(\beta)} = e_{\underline{f}(\beta)}$. Consider $x_\alpha \in S_\alpha \subseteq S,\ \alpha \in Y$, we have

$$f_\beta(\varphi_{\alpha,\beta}(x_\alpha)) = f_\beta(e_\beta) = e_{\underline{f}(\beta)} = \varphi_{s(\alpha),\underline{f}(\beta)}(f_\alpha(x_\alpha))$$

Thus $f_\beta \varphi_{\alpha,\beta} = \varphi_{\underline{f}(\alpha),\underline{f}(\beta)} f_\alpha$. □

The next construction is useful and often used later.

Construction 3.1.1. *Let $S = [Y; S_\alpha, e_\alpha, \varphi_{\alpha,\beta}]$, $\varphi_{\alpha,\beta} = c_{e_\beta}$ be a non-trivial strong semilattice of left simple semigroups with $\nu = \wedge Y$. Take $f_\alpha \in Hom(S_\alpha, S_\beta)$, $\alpha, \beta \in Y$. Define $f : S \to S$ as follows*

$$f(x_\xi) := \begin{cases} f_\alpha(x_\alpha) \in S_\beta & \text{if } \alpha = \xi, \\ f_\alpha(e_\alpha) \in S_\beta & \text{if } \alpha < \xi, \\ e_\nu & \text{if } \xi < \alpha \text{ or } \alpha \| \xi, \end{cases}$$

for every $x_\xi, \in S,\ \xi \in Y$. Then $f \in End(S)$.

Proof. It can be seen that f is well-defined. We check that f is a homomorphism.

Take $x_\gamma, y_\delta \in S,\ \gamma, \delta \in Y$.

The case $\gamma = \delta = \alpha$ is clear as $f_\alpha \in Hom(S_\alpha, S_\beta)$.

Case 1.1. $\gamma = \alpha,\ \alpha < \delta$. Then $\alpha = \alpha\delta$. We calculate

$$\begin{aligned} f(x_\alpha y_\delta) &= f(x_\alpha \varphi_{\delta,\alpha}(y_\delta)) \\ &= f(x_\alpha e_\alpha) \\ &= f(x_\alpha) \\ &= f_\alpha(x_\alpha) \end{aligned}$$

and

$$\begin{aligned} f(x_\alpha)f(y_\delta) &= f_\alpha(x_\alpha)f_\alpha(e_\alpha) \\ &= f_\alpha(x_\alpha e_\alpha) \\ &= f_\alpha(x_\alpha). \end{aligned}$$

Case 1.2. $\gamma = \alpha,\ \delta < \alpha$ or $\delta \| \alpha$. Then $\alpha\delta < \alpha$. We calculate

$$\begin{aligned} f(x_\alpha y_\delta) &= f(\varphi_{\alpha,\alpha\delta}(x_\alpha)\varphi_{\delta,\alpha\delta}(y_\delta)) \\ &= f(e_{\alpha\delta}) \\ &= e_\nu \end{aligned}$$

and

$$\begin{aligned} f(x_\alpha)f(y_\delta) &= f_\alpha(x_\alpha)e_\nu \\ &= \varphi_{\beta,\nu}(f_\alpha(x_\alpha))e_\nu \\ &= e_\nu. \end{aligned}$$

Case 1.3. $\alpha < \gamma, \delta$. Then $\alpha \leq \gamma\delta$.

If $\alpha = \gamma\delta$ then

$$\begin{aligned} f(x_\gamma y_\delta) &= f(\varphi_{\gamma,\alpha}(x_\gamma)\varphi_{\delta,\alpha}(y_\delta)) \\ &= f(e_\alpha) \\ &= f_\alpha(e_\alpha) \end{aligned}$$

and

$$\begin{aligned} f(x_\gamma)f(y_\delta) &= f_\alpha(e_\alpha)f_\alpha(e_\alpha) \\ &= f_\alpha(e_\alpha e_\alpha) \\ &= f_\alpha(e_\alpha). \end{aligned}$$

If $\alpha < \gamma\delta$ then

$$\begin{aligned} f(x_\gamma y_\delta) &= f(\varphi_{\gamma,\gamma\delta}(x_\gamma)\varphi_{\delta,\gamma\delta}(y_\delta)) \\ &= f(e_{\gamma\delta}) \\ &= f_\alpha(e_\alpha) \end{aligned}$$

and

$$\begin{aligned} f(x_\gamma)f(y_\delta) &= f_\alpha(e_\alpha)f_\alpha(e_\alpha) \\ &= f_\alpha(e_\alpha e_\alpha) \\ &= f_\alpha(e_\alpha). \end{aligned}$$

Case 1.4. $\alpha < \gamma$ and ($\delta < \alpha$ or $\delta \| \alpha$).

If $\alpha < \gamma$ and $\delta < \alpha$ then $\delta = \delta\gamma$. We calculate

$$\begin{aligned} f(x_\gamma y_\delta) &= f(\varphi_{\gamma,\delta}(x_\gamma)y_\delta) \\ &= f(e_\delta y_\delta) \\ &= e_\nu \end{aligned}$$

and

$$\begin{aligned} f(x_\gamma)f(y_\delta) &= f_\alpha(e_\alpha)e_\nu \\ &= \varphi_{\beta,\nu}(f_\alpha(e_\alpha))e_\nu \\ &= e_\nu. \end{aligned}$$

If $\alpha < \gamma$ and $\delta \| \alpha$ then $\alpha \neq \gamma\delta$ since otherwise $\alpha = \gamma\delta < \delta$ but $\alpha \| \delta$. Moreover $\alpha \not< \gamma\delta$ since otherwise $\alpha < \gamma\delta < \delta$ but $\alpha \| \delta$. We have

$$\begin{aligned} f(x_\gamma y_\delta) &= f(\varphi_{\gamma,\gamma\delta}(x_\gamma)\varphi_{\delta,\gamma\delta}(y_\delta)) \\ &= f(e_{\gamma\delta}) \\ &= e_\nu \end{aligned}$$

and

$$\begin{aligned} f(x_\gamma)f(y_\delta) &= f_\alpha(e_\alpha)e_\nu \\ &= \varphi_{\beta,\nu}(f_\alpha(e_\alpha))e_\nu \\ &= e_\nu. \end{aligned}$$

Case 1.5. ($\gamma < \alpha$ or $\gamma \| \alpha$) and ($\delta < \alpha$ or $\delta \| \alpha$).

If $\gamma < \alpha$ and $\delta < \alpha$ then $\gamma\delta < \alpha$. We calculate

$$\begin{aligned} f(x_\gamma y_\delta) &= f(\varphi_{\gamma,\gamma\delta}(x_\gamma)\varphi_{\delta,\gamma\delta}(y_\delta)) \\ &= f(e_{\gamma\delta}) \\ &= e_\nu \\ &= f(x_\gamma)f(y_\delta). \end{aligned}$$

If $\gamma < \alpha$ and $\delta \| \alpha$ then $\gamma\delta < \gamma < \alpha$. We calculate

$$
\begin{aligned}
f(x_\gamma y_\delta) &= f(\varphi_{\gamma,\gamma\delta}(x_\gamma)\varphi_{\delta,\gamma\delta}(y_\delta)) \\
&= f(e_{\gamma\delta}) \\
&= e_\nu \\
&= f(x_\gamma)f(y_\delta).
\end{aligned}
$$

If $\gamma \| \alpha$ and $\delta \| \alpha$ then $\gamma\delta \neq \alpha$ since otherwise $\alpha = \gamma\delta < \gamma$ but $\alpha \| \gamma$. Moreover $\alpha \not< \gamma\delta$ since otherwise $\alpha < \gamma\delta < \delta$ but $\alpha \| \delta$. We have

$$
\begin{aligned}
f(x_\gamma y_\delta) &= f(\varphi_{\gamma,\gamma\delta}(x_\gamma)\varphi_{\delta,\gamma\delta}(y_\delta)) \\
&= f(e_{\gamma\delta}) \\
&= e_\nu \\
&= f(x_\gamma)f(y_\delta).
\end{aligned}
$$

Thus $f \in End(S)$. □

Now we consider the case that the defining homomorphisms $\varphi_{\alpha,\beta}$ are isomorphisms, some places the author write as bijective. In this case we simplify the description of a non-trivial strong semilattice of left simple semigroups $S = [Y; T_\alpha, e_\alpha, \varphi_{\alpha,\beta}]$ such that the result covers Lemma 2.2 of Gilbert and Samman [4].

For $\alpha, \beta \in Y$, $T_\alpha \cong T_\beta$ and $\varphi_{\alpha,\beta}(e_\alpha) = e_\beta$ can be taken without loss of generality.

Lemma 3.1.5. *Let $S = [Y; T_\alpha, e_\alpha, \varphi_{\alpha,\beta}]$ be a non-trivial strong semilattice of semigroups with isomorphisms $\varphi_{\alpha,\beta}$. For any $\lambda \in Y$, let $S_\lambda = [Y; T_\lambda, e_\alpha, id_{\alpha,\alpha}]$ be the strong semilattice of semigroups over Y in which each semigroups T_α, $\alpha \in Y$ is equal to T_λ and all the defining homomorphisms are the identity. Then S is isomorphic to S_λ.*

Proof. We define an isomorphism $\phi : S \to S_\lambda$ as follows

$$\phi(a) := \phi_\alpha(a) = \varphi^{-1}_{\lambda,\alpha\lambda}\varphi_{\alpha,\alpha\lambda}(a)$$

for every $a \in T_\alpha \subseteq S$ where ϕ_α is the restriction to T_α. Then ϕ is clearly bijective and we check only that ϕ is a homomorphism.

Let $a \in T_\alpha, b \in T_\beta$. Then $ab = \varphi_{\alpha,\alpha\beta}(a)\varphi_{\beta,\alpha\beta}(b) \in T_{\alpha\beta}$ and

$$
\begin{aligned}
\phi(a)\phi(b) &= \phi_\alpha(a)\phi_\beta(b) \\
&= \varphi^{-1}_{\lambda,\alpha\lambda}\varphi_{\alpha,\alpha\lambda}(a)\varphi^{-1}_{\lambda,\beta\lambda}\varphi_{\beta,\beta\lambda}(b)
\end{aligned}
$$

whereas

$$
\begin{aligned}
\phi(ab) &= \phi_{\alpha\beta}(\varphi_{\alpha,\alpha\beta}(a)\varphi_{\beta,\alpha\beta}(b)) \\
&= \phi_{\alpha\beta}(\varphi_{\alpha,\alpha\beta}(a))\phi_{\alpha\beta}(\varphi_{\beta,\alpha\beta}(b)).
\end{aligned}
$$

Consider

$$\begin{aligned}\phi_{\alpha\beta}(\varphi_{\alpha,\alpha\beta}(a)) &= \varphi^{-1}_{\lambda,\alpha\beta\lambda}\varphi_{\alpha\beta,\alpha\beta\lambda}(\varphi_{\alpha,\alpha\beta}(a))\\ &= (\varphi^{-1}_{\lambda,\alpha\lambda}\varphi^{-1}_{\alpha\lambda,\alpha\beta\lambda})(\varphi_{\alpha\lambda,\alpha\beta\lambda}\varphi_{\alpha,\alpha\lambda})(a)\\ &= \varphi^{-1}_{\lambda,\alpha\lambda}\varphi_{\alpha,\alpha\lambda}(a).\end{aligned}$$

Similarly, $\phi_{\alpha\beta}(\varphi_{\beta,\alpha\beta}(b)) = \varphi^{-1}_{\lambda,\beta\lambda}\varphi_{\beta,\beta\lambda}(b)$. Thus ϕ is a homomorphism. □

As we know from Lemma 3.1.5 that the results are similar for the defining homomorphisms being isomorphisms or identity, so from now on we prove the latter case, but we write isomorphisms instead.

We now assume that S is a non-trivial strong semilattice of left simple semigroups over Y in which every left simple semigroup is equal to a fixed left simple semigroup T and with each defining homomorphism equal to the identity. Hence S is the disjoint union of copies T_α (i.e., T indexed by $\alpha \in Y$). If $x \in T$, then we denote by x_α the copy of the element x in T_α. Thus

$$x_\alpha y_\beta = (xy)_{\alpha\beta}.$$

Construction 3.1.2. *Let* $S = [Y; T_\alpha, e_\alpha, \varphi_{\alpha,\beta}]$ *be a non-trivial strong semilattice of semigroups with isomorphisms* $\varphi_{\alpha,\beta}$, *i.e.,* $T_\alpha \cong T_\beta \cong T$. *Any* $g \in End(T)$ *and* $s \in End(Y)$ *determine an endomorphism* $f \in End(S)$ *defined by* $f(x_\alpha) := (g(x))_{s(\alpha)}$.

Proof. It can be seen that f is well-defined.

Take $x_\alpha, y_\beta \in S$. Then

$$\begin{aligned}f(x_\alpha y_\beta) &= f(\varphi_{\alpha,\alpha\beta}(x_\alpha)\varphi_{\beta,\alpha\beta}(y_\beta))\\ &= f(x_{\alpha\beta}y_{\alpha\beta})\\ &= (g(xy))_{s(\alpha\beta)}\end{aligned}$$

and

$$\begin{aligned}f(x_\alpha)f(y_\beta) &= (g(x))_{s(\alpha)}(g(y))_{s(\beta)}\\ &= \varphi_{s(\alpha),s(\alpha)s(\beta)}((g(x))_{s(\alpha)})\varphi_{s(\beta),s(\alpha)s(\beta)}((g(y))_{s(\beta)})\\ &= (g(x))_{s(\alpha)s(\beta)}(g(y))_{s(\alpha)s(\beta)} \in T_{s(\alpha\beta)}\\ &= (g(xy))_{s(\alpha\beta)}.\end{aligned}$$

Since $s \in End(Y)$ and $g \in End(T)$, we have $f(x_\alpha y_\beta) = f(x_\alpha)f(y_\beta)$. Hence $f \in End(S)$. □

Proposition 3.1.1. *Let $S = [Y; T_\alpha, e_\alpha, \varphi_{\alpha,\beta}]$ be a non-trivial strong semilattice of left simple semigroups with isomorphisms $\varphi_{\alpha,\beta}$, i.e., $T_\alpha \cong T_\beta \cong T$. Take $f \in End(S)$. Then there exists $g \in End(T) = End(T_\alpha)$ and $\underline{f} \in End(Y)$ such that $f(x_\alpha) = (g(x))_{\underline{f}(\alpha)}$.*

Proof. Since $f \in End(S)$, we have $\underline{f} \in End(Y)$ by Corollary 3.1.1. For each $x_\alpha \in S$, $\alpha \in Y$, we have $f_\alpha \in Hom(T_\alpha, T_{\underline{f}(\alpha)})$, but $T_\alpha = T_{\underline{f}(\alpha)} = T$. Then there exists $g \in End(T)$ such that $f_\alpha = g$ and

$$f(x_\alpha) = (g(x))_{\underline{f}(\alpha)}.$$

□

The following theorem is a consequence of Construction 3.1.2 and Proposition 3.1.1.

Theorem 3.1.1. *Let $S = [Y; T_\alpha, e_\alpha, \varphi_{\alpha,\beta}]$ be a non-trivial strong semilattice of left simple semigroups T_α with isomorphisms $\varphi_{\alpha,\beta}$, i.e., $T_\alpha \cong T_\beta \cong T$. Every endomorphism is of the forms such that $f \in End(S)$ if and only if there exist $g \in End(T)$ and $s \in End(Y)$ with $f(x_\alpha) = (g(x))_{s(\alpha)}$ and $\underline{f}(\alpha) = s(\alpha)$ for every $x_\alpha \in S$, $\alpha \in Y$.*

Proof. See Construction 3.1.2 and Proposition 3.1.1. □

3.2 Regular monoids

In this section we consider strong semilattices of left simple semigroups whose endomorphism monoids are regular.

Lemma 3.2.1. *Let $S = [Y; S_\alpha, e_\alpha, c_{\alpha,e_\beta}]$ be a non-trivial strong semilattice of semigroups. If the monoid $End(S)$ is regular then the set $Hom(S_\alpha, S_\beta)$ is hom-regular for all $\alpha \in Y$.*

Proof. Take $f_\alpha \in Hom(S_\alpha, S_\beta)$, $\alpha, \beta \in Y$. Using Construction 3.1.1, for every $x_\xi, \in S$, $\xi \in Y$, take $f \in End(S)$ as follows

$$f(x_\xi) := \begin{cases} f_\alpha(x_\alpha) \in S_\beta & \text{if } \alpha = \xi, \\ f_\alpha(e_\alpha) \in S_\beta & \text{if } \alpha < \xi, \\ e_\nu & \text{if } \xi < \alpha \text{ or } \alpha \| \xi, \end{cases}$$

By hypothesis there exists $f' \in End(S)$ such that $ff'f = f$.

For each $x_\alpha \in S,\ \alpha \in Y$. We calculate

$$\begin{aligned} f_\alpha(x_\alpha) &= f(x_\alpha) \\ &= ff'f(x_\alpha) \\ &= ff'(f_\alpha(x_\alpha)) \\ &= f_\gamma f'_\beta f_\alpha(x_\alpha) \end{aligned}$$

where $f'_\beta \in Hom(S_\beta, S_\gamma)$ for some $\gamma \in \underline{f}^{-1}\{\beta\}$ and $f_\gamma \in Hom(S_\gamma, S_\beta)$.

If $\alpha < \gamma$ then $f_\alpha(x_\alpha) = f_\gamma(f'_\beta f_\alpha(x_\alpha)) = f_\alpha(e_\alpha)$, i.e., f_α is constant, of course f is regular.

If $\gamma \| \alpha$ or $\gamma < \alpha$ then $f_\alpha(x_\alpha) = f_\gamma(f'_\beta f_\alpha(x_\alpha)) = e_\nu$, i.e., f_α is constant onto e_ν, of course f is regular.

If $\gamma = \alpha$ then $f_\alpha(x_\alpha) = f_\alpha f'_\beta f_\alpha(x_\alpha)$, i.e., f_α is regular. □

If we take $\alpha = \beta$ in Construction 3.1.1, we have the following lemma.

Lemma 3.2.2. *Let $S = [Y; S_\alpha, e_\alpha, c_{\alpha,e_\beta}]$ be a non-trivial strong semilattice of semigroups. If the monoid End(S) is regular (idempotent-closed, orthodox, left inverse, completely regular, and idempotent), then the monoid $End(S_\alpha)$ is regular (idempotent-closed, orthodox, left inverse, completely regular, and idempotent).*

Lemma 3.2.3. *Let $S = [Y; S_\alpha, e_\alpha, \varphi_{\alpha,\beta}]$ be a non-trivial strong semilattice of semigroups. If the monoid End(S) is regular (idempotent-closed, orthodox, left inverse, completely regular, and idempotent), then the monoid End(Y) is regular (idempotent-closed, orthodox, left inverse, completely regular, and idempotent).*

Proof. Take $s \in End(Y)$. Using Lemma 3.1.2, take $f \in End(S)$ as follows

$$f(x_\alpha) := e_{s(\alpha)}$$

for every $x_\alpha \in S,\ \alpha \in Y$. By hypothesis there exists $f' \in End(S)$ such that $ff'f = f$. Further we have $e_{s(\alpha)} = f(x_\alpha) = ff'(f(x_\alpha)) = ff'(e_{s(\alpha)}) = e_{s\underline{f}s(\alpha)}$. So that $s(\alpha) = s\underline{f}s(\alpha)$, and therefore s is regular. Hence the monoid End(Y) is regular.

The remaining properties can be proved in a similar way. □

Lemma 3.2.4. *Let $S = [Y; S_\alpha, e_\alpha, c_{\alpha,e_\beta}]$ be a non-trivial strong semilattice of semigroups with $\nu = \wedge Y$. If the monoid End(S) is regular, then the set $Hom(S_\nu, S_\alpha)$ consists of constant maps for every $\alpha \in Y,\ \alpha \neq \nu$.*

Proof. Take $f_\nu \in Hom(S_\nu, S_\alpha)$. For each $x_\nu \in S_\nu$ we know that $f_\nu(x_\nu) \in S_\alpha$ then $f_\nu(x_\nu) = y_\alpha$ for some $y_\alpha \in S_\alpha$. Take $s \in End(Y)$ such that $s(\xi) = \alpha$ for all $\xi \in Y$. Define $f \in End(S)$ as follows

$$f(z_\xi) := \begin{cases} f_\nu(z_\nu) \in S_\alpha & \text{if } \xi = \nu, \\ f_\nu(e_\nu) \in S_\alpha & \text{otherwise,} \end{cases}$$

for every $z_\xi \in S, \xi \in Y$. For $\beta, \gamma \in Y$, consider

$$f_\beta \varphi_{\gamma,\beta}(z_\gamma) = f_\beta(e_\beta) = f_\nu(e_\nu)$$

and

$$\varphi_{s(\gamma),s(\beta)}(f_\gamma(z_\gamma)) = \varphi_{\alpha,\alpha}(f_\gamma(z_\gamma)) = f_\gamma(z_\gamma) = f_\nu(e_\nu).$$

Thus the set $\{f_\beta \in Hom(S_\beta, S_{s(\beta)}) \mid \beta \in Y\}$ satisfies the equations

$$f_\beta \varphi_{\gamma,\beta} = \varphi_{s(\gamma),s(\beta)} f_\gamma.$$

Then $f \in End(S)$ by Lemma 3.1.2. By hypothesis there exists $f' \in End(S)$ such that $ff'f = f$. Then

$$\begin{aligned} y_\alpha &= f_\nu(x_\nu) \\ &= f(x_\nu) \\ &= ff'f(x_\nu) \\ &= ff'(f_\nu(x_\nu)) \\ &= ff'(y_\alpha). \end{aligned}$$

Thus

$$f'(y_\alpha e_\nu) = f'(\varphi_{\alpha,\nu}(y_\alpha)e_\nu) = f'(e_\nu e_\nu) = f'(e_\nu)$$

and

$$f'(y_\alpha)f'(e_\nu) = f'(y_\alpha).$$

That is $f'(y_\alpha) = f'(e_\nu)$.

Since $f'(y_\alpha)$ must be in S_ν because $ff'(y_\alpha) = y_\alpha$ and by Lemma 2.2 $f'(S_\alpha) \subseteq S_\nu$, i.e., $\underline{f}'(\alpha) = \nu$, so that

$$\underline{f}'(\nu) = \underline{f}'(\nu\alpha) = \underline{f}'(\nu)\underline{f}'(\alpha) = \underline{f}'(\nu)\nu = \nu.$$

Now we claim that $f'(S_\alpha) = \{f'(e_\nu)\}$. Take any $z_\alpha \in S_\alpha$. Since $f' \in End(S)$, we have

$$f'_\nu(\varphi_{\alpha,\nu}(z_\alpha)) = f'_\nu(e_\nu) = f'(e_\nu)$$

and

$$\varphi_{\underline{f'}(\alpha),\underline{f'}(\nu)}(f'_\alpha(z_\alpha)) = \varphi_{\nu,\nu}(f'_\alpha(z_\alpha)) = f'_\alpha(z_\alpha) = f'(z_\alpha).$$

We get that $f'(S_\alpha) = \{f'(e_\nu)\}$. Then

$$\begin{aligned} f_\nu(e_\nu) &= f(e_\nu) \\ &= ff'f(e_\nu) \\ &= ff'(e'_\alpha) \text{ where } f(e_\nu) = e'_\alpha \text{ for some } e'_\alpha \in S_\alpha \\ &= ff'(e_\nu) \text{ (because } f'(S_\alpha) = \{f'(e_\nu)\}) \\ &= ff'(y_\alpha) \text{ (because } f'(y_\alpha) = f'(e_\nu)) \\ &= y_\alpha \\ &= f_\nu(x_\nu). \end{aligned}$$

This implies that f_ν is constant onto $f_\nu(e_\nu)$. Therefore every $f_\nu \in Hom(S_\nu, S_\alpha)$ is a constant mapping. Hence $Hom(S_\nu, S_\alpha)$ consists of constant maps for all $\alpha \in Y$, $\alpha > \nu$. □

The following lemma is needed later.

Lemma 3.2.5. *Let $Y = Y_{0,n}$ and $S = [Y_{0,n}; S_\alpha, e_\alpha, c_{\alpha,e_\beta}]$ be a non-trivial strong semilattice of left simple semigroups. Take $f \in End(S)$. If $\underline{f}(\xi) = \alpha$ for all $\xi \in Y$ for some $0 \neq \alpha \in Y_{0,n}$, then $f(x_\beta) = f(e_0)$ for all $0 \neq \beta \in Y_{0,n}$. Moreover, f is idempotent if and only if $f(S) = f(e_0)$. In fact, f is constant onto $f(e_0)$.*

Proof. Take $x_\beta \in S_\beta$, $\beta \in Y$.

Since $f \in End(S)$, we have $\underline{f} \in End(Y_{0,n})$ by Corollary 3.1.1 and the set $\{f_\alpha \in Hom(S_\alpha, S_{\underline{f}(\alpha)}) \mid \alpha \in Y_{0,n}\}$ satisfies the equations

$$f_0\varphi_{\beta,0} = \varphi_{\underline{f}(\beta),\underline{f}(0)}f_\beta$$

we have

$$\begin{aligned} f(x_\beta) &= f_\beta(x_\beta) \\ &= \varphi_{\underline{f}(\beta),\underline{f}(0)}f_\beta(x_\beta) \text{ (because } \underline{f}(0) = \underline{f}(\beta)) \\ &= f_0(\varphi_{\beta,0}(x_\beta)) \\ &= f_0(e_0) \\ &= f(e_0). \end{aligned}$$

Thus $f(x_\beta) = f(e_0)$ for $x_\beta \in S_\beta$.

Moreover, let f be idempotent. We have shown from above that $f(x_\beta) = f(e_0)$ for every $0 \neq \beta \in Y_{0,n}$. So that we need only show that $f(S_0) = f(e_0)$. Take $x_0 \in S_0$.

Then $f(x_0) = f(f(x_0)) = f(y_\alpha)$ for some $y_\alpha \in S_\alpha$. But $f(y_\alpha) = f(e_0)$ from above. Thus $f(x) = f(y_\alpha) = f(e_0)$ for all $x \in S$.

On the other hand, the image $f(S) = f(e_0)$ is idempotent, and therefore f is idempotent. □

Now we proceed to a sufficient condition for the case $Y = Y_{0,n}$ i.e., Y has only one $\wedge$-reducible element. We have.

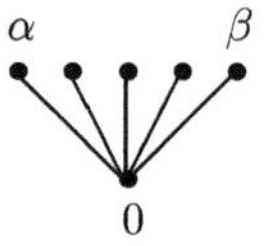

Theorem 3.2.1. *Let $S = [Y; S_\alpha, e_\alpha, c_{\alpha,e_\beta}]$ be a non-trivial strong semilattice of left simple semigroups. If the following hold*

1) $Y = Y_{0,n}$,

2) the set $Hom(S_0, S_\alpha)$ consists of constant mappings for all $\alpha \in Y_{0,n}$, $\alpha \neq 0$,

3) the set $Hom(S_\alpha, S_\beta)$ is hom-regular for all $\alpha, \beta \in Y_{0,n}$, and

4) S_0 contains only one idempotent e_0,

then the monoid $End(S)$ is regular.

Proof. Take $f \in End(S)$. Then $\underline{f} \in End(Y_{0,n})$.

Case 1. $\underline{f}$ is constant.

If $\underline{f}(\xi) = 0$ for all $\xi \in Y_{0,n}$, then $f(S_\alpha) = \{f(e_0)\}$ for all $0 \neq \alpha \in Y_{0,n}$ by Lemma 3.2.4. Thus f is determined by $f_0 \in End(S_0)$, so that $f|_{S_0} = f_0$ and $f|_{S_\alpha} = f_0\varphi_{\alpha,0}$, that is $f_\alpha(x_\alpha) = f_0(\varphi_{\alpha,0}(x_\alpha)) = f_0(e_0)$ for every $x_\alpha \in S_\alpha$. By using that $End(S_0)$ is regular, we get that f is regular.

If $\underline{f}(\xi) = \alpha$ for all $\xi \in Y_{0,n}$ and some $0 \neq \alpha \in Y_{0,n}$, then f is determined by $f_0 \in Hom(S_0, S_\alpha)$ and $f|_{S_0} = f_0$ and $f|_{S_\alpha} = f_0\varphi_{\alpha,0}$, that is $f_\alpha(x_\alpha) = f_0(\varphi_{\alpha,0}(x_\alpha)) = f_0(e_0)$ for every $x_\alpha \in S_\alpha$. By using that 1) $Hom(S_0, S_\alpha)$ consists of constant maps, we get that f is regular.

Case 2. $\underline{f}$ is not constant.

If $\underline{f}(0) = \alpha$ for some $0 \neq \alpha \in Y_{0,n}$ then for every $0 \neq \beta, \gamma \in Y_{0,n}$ we have $\underline{f}(\beta)\underline{f}(\gamma) = \underline{f}(\beta\gamma) = \underline{f}(0) = \alpha$. This implies that $\underline{f}(\beta) = \alpha$ and $\underline{f}(\gamma) = \alpha$ which is impossible as $\underline{f}$ is not constant, so that $\underline{f}(0) = 0$.

Now consider any $\alpha \in Y_{0,n}$ with $\underline{f}(\alpha) \neq 0$, say $\underline{f}(\alpha) = \beta$ for some $0 \neq \beta \in Y_{0,n}$. Then f is determined by each $f_\alpha \in Hom(S_\alpha, S_\beta)$ and $f_0 \in End(S_0)$ such that $f|_{S_0} = f_0$ and $f_0\varphi_{\alpha,0} = \varphi_{\underline{f}(\alpha),0}f_\alpha$, i.e.,

$$f_0(e_0) = f_0(\varphi_{\alpha,0}(x_\alpha)) = \varphi_{\beta,0}(f_\alpha(x_\alpha)) = e_0.$$

If there exists $\gamma \in Y_{0,n}$ such that $\underline{f}(\gamma) = \beta$ then $\beta = \underline{f}(\gamma)\underline{f}(\alpha) = \underline{f}(\gamma\alpha) = \underline{f}(0) = 0$, but $\beta \neq 0$. This means that $\underline{f}^{-1}\{\beta\} = \{\alpha\}$ and for any two distinct elements $0 \neq \gamma, \delta \in Y_{0,n}$ such that $\underline{f}(\gamma) = \underline{f}(\delta) = \eta$ then $\eta = 0$, i.e., $|\underline{f}^{-1}(\beta)| = 1$ (the cardinality of $\underline{f}^{-1}(\beta)$). By using 2) there exists $f'_\beta \in Hom(S_\beta, S_\alpha)$ such that $f_\alpha f'_\beta f_\alpha = f_\alpha$ while $f'_0 \in End(S_0)$ with $f'(e_0) = e_0$ since S_0 has only one idempotent e_0 by 3).

Then $f' \in End(S)$ can be defined by using Lemma 3.1.2

$$f'(x_\xi) := \begin{cases} f'_\xi(x_\xi) & \text{if } \xi \in Im(\underline{f}), \\ e_0 & \text{otherwise}, \end{cases}$$

for every $x_\xi \in S$, $\xi \in Y_{0,n}$, and

$$\begin{aligned} ff'f(x_\alpha) &= ff'(f_\alpha(x_\alpha)) \\ &= ff'_\beta(f_\alpha(x_\alpha)) \text{ where } \underline{f}(\alpha) = \beta \\ &= f_\alpha f'_\beta f_\alpha(x_\alpha) \\ &= f_\alpha(x_\alpha) \\ &= f(x_\alpha). \end{aligned}$$

Thus f is regular. □

The following lemma will be used later.

We recall that

$$[\alpha) = \{\beta \in Y \mid \beta \geq \alpha\}.$$

Lemma 3.2.6. *Let $S = [Y; S_\alpha, e_\alpha, c_{\alpha,e_\beta}]$ be a non-trivial strong semilattice of left simple semigroups. Take $f \in End(S)$. If $\underline{f}(\alpha) = \underline{f}(\beta)$, then $f(x_\beta) = f(e_\alpha)$ for all $x_\beta \in S_\beta$, $\beta \in [\alpha)\backslash\{\alpha\}$*

Proof. Since $\beta \in [\alpha)$, we have $\varphi_{\beta,\alpha}(x_\beta) = e_\alpha$. Then

$$\begin{aligned} f(x_\beta) &= f_\beta(x_\beta) \\ &= \varphi_{\underline{f}(\beta),\underline{f}(\alpha)}(f_\beta(x_\beta)) \\ &= f_\alpha(\varphi_{\beta,\alpha}(x_\beta)) \\ &= f_\alpha(e_\alpha) \\ &= f(e_\alpha). \end{aligned}$$

Therefore $f(x_\beta) = f(e_\alpha)$. □

From Lemma 3.2.6 we note that for $\beta > \alpha$ and both are sent to the same image, the homomorphism from S_β to $S_{\underline{f}(\beta)}$ will be constant onto $f(e_\alpha)$.

Theorem 3.2.2. *Let $S = [Y; S_\alpha, e_\alpha, c_{\alpha,e_\beta}]$ be a non-trivial strong semilattice of left simple semigroups with $\nu = \wedge Y$. If the monoid End(S) is regular then the following conditions hold*

1) End(Y) is regular, i.e., Y is a binary tree or $Y = Y_{0,n}$ or $Y \in \mathbf{B} \cup \mathbf{B}^d \cup \mathbf{R}$ (see Theorem 2.1.1),

2) the set $Hom(S_\nu, S_\alpha)$ consists of constant mappings for all $\alpha \in Y$, $\nu < \alpha$, and

3) the set $Hom(S_\alpha, S_\beta)$ is hom-regular for every $\alpha, \beta \in Y$.

If all defining homomorphisms $\varphi_{\alpha,\beta}$ are isomorphisms we have.

Theorem 3.2.3. *Let $S = [Y; T_\alpha, e_\alpha, \varphi_{\alpha,\beta}]$ be a non-trivial strong semilattice of left simple semigroups with isomorphisms $\varphi_{\alpha,\beta}$, i.e., $T_\alpha \cong T_\beta \cong T$. Then the monoid End(S) is regular if and only if the following assertions hold*

1) the monoid End(Y) is regular,

2) the monoid End(T) is regular.

Proof. Necessity. 1) follows from Lemma 3.2.3.

2) We now show that *End(T)* is regular. Take $g \in End(T)$. Using Construction 3.1.2, take $f \in End(S)$ as follows

$$f(x_\alpha) := (g(x))_\alpha$$

for every $x_\alpha \in S$, $\alpha \in Y$. It can be seen that $\underline{f}(\alpha) = \alpha$ for all $\alpha \in Y$. By hypothesis there exists $f' \in End(S)$ such that $ff'f = f$. We set $(g'(x))_\alpha = f'(x_\alpha)$ for all $x_\alpha = x \in T$, $\alpha \in Y$ such that $(gg'g(x))_\alpha = gg'((g(x))_\alpha) = gg'f(x_\alpha) = g(f'f(x_\alpha)) = (ff'f)(x_\alpha) = f(x_\alpha) = (g(x))_\alpha$ and $g \in End(T_\alpha) = End(T)$. Therefore *End(T)* is regular.

Sufficiency. Assume that *End(Y)* and *End(T)* are regular. Take $f \in End(S)$. Then $\underline{f} \in End(Y)$ and in fact, $f|_{T_\alpha} : T_\alpha \to T_{\underline{f}(\alpha)} = g$ for some $g \in End(T)$. By assumption, there exist $g' \in End(T)$ and $s \in End(Y)$ such that $\underline{f}\ s\ \underline{f} = \underline{f}$ and $gg'g = g$. Using Construction 3.1.2, take $f' \in End(S)$ as follows

$$f'(x_\alpha) := (g'(x))_{s(\alpha)}$$

for every $x_\alpha \in S,\ \alpha \in Y$ such that $ff'f = f$. Thus f is regular, and therefore $End(S)$ is regular. □

Remark 3.2.1. All the results in this chapter hold for the strong semilattices of right simple semigroups as well.

Example 3.2.1. left zero semigroups are left simple semigroups and the endomorphism monoids of left zero semigroups are regular since the set of endomorphism monoids of a left zero semigroup is isomorphic to the set of transformations of a set. For others properties see also Corollary 2.2.3.

Problem 3.2.1. Investigate left simple semigroups S with regular endomorphism monoid

3.3 Idempotent-closed monoids

In this section we consider the strong semilattices of left simple semigroups whose endomorphism monoids are idempotent-closed.

Construction 3.3.1. *Let $S = [Y; S_\alpha, e_\alpha, \varphi_{\alpha,\beta}]$ be a non-trivial strong semilattice of semigroups with $\nu = \wedge Y$. Take $f_\nu \in End(S_\nu)$. Define $f \in End(S)$ as follows*

$$f(x_\xi) := \begin{cases} f_\nu(x_\nu) & \text{if } \xi = \nu, \\ f_\nu(\varphi_{\xi,\nu}(x_\xi)) & \text{if } \xi \neq \nu, \end{cases}$$

for every $x_\xi \in S,\ \xi \in Y$. Then $f \in End(S)$.

Proof. It can be seen that f is well-defined. Now we show that f is a homomorphism.

Take $x_\alpha, y_\beta \in S,\ \alpha, \beta \in Y$. Then

$$\begin{aligned} f(x_\alpha y_\beta) &= f(\varphi_{\alpha,\alpha\beta}(x_\alpha)\varphi_{\beta,\alpha\beta}(y_\beta)) \\ &= f_\nu(\varphi_{\alpha\beta,\nu}(\varphi_{\alpha,\alpha\beta}(x_\alpha)\varphi_{\beta,\alpha\beta}(y_\beta))) \\ &= f_\nu((\varphi_{\alpha\beta,\nu}\varphi_{\alpha,\alpha\beta}(x_\alpha))(\varphi_{\alpha\beta,\nu}\varphi_{\beta,\alpha\beta}(y_\beta))) \\ &= f_\nu(\varphi_{\alpha,\nu}(x_\alpha))f_\nu(\varphi_{\beta,\nu}(y_\beta)) \\ &= f(x_\alpha)f(y_\beta). \end{aligned}$$

Therefore $f \in End(S)$. □

Lemma 3.3.1. *Let $S = [Y; S_\alpha, e_\alpha, c_{\alpha,e_\beta}]$ be a non-trivial strong semilattice of semigroups. If the monoid End(S) is idempotent-closed, then $Y = Y_{0,n}$ and the monoid $End(S_\xi)$ is idempotent-closed for every $\xi \in Y_{0,n}$.*

Proof. From Lemma 3.2.3 we get that $End(Y)$ is idempotent-closed and the monoid $End(Y)$ is idempotent-closed if and only if $Y = Y_{0,n}$ by Proposition 2.2.1.

We verify first that $End(S_0)$ is idempotent-closed. Take two idempotents $f_0, h_0 \in End(S_0)$. Using Construction 3.3.1, for every $x_\xi \in S$, $\xi \in Y_{0,n}$, take $f, h \in End(S)$ as follows

$$f(x_\xi) := \begin{cases} f_0(x_0) & \text{if } \xi = 0, \\ f_0(\varphi_{\xi,0}(x_\xi)) & \text{if } \xi \neq 0, \end{cases}$$

and

$$h(x_\xi) := \begin{cases} h_0(x_0) & \text{if } \xi = 0, \\ h_0(\varphi_{\xi,0}(x_\xi)) & \text{if } \xi \neq 0. \end{cases}$$

Then f, h are idempotents. By hypothesis fh is idempotent. Then

$$\begin{aligned} f_0h_0f_0h_0(x_0) &= fhfh(x_0) \\ &= fh(x_0) \\ &= f_0h_0(x_0). \end{aligned}$$

Thus f_0h_0 is idempotent, and therefore $End(S_0)$ is idempotent-closed.

We now show that $End(S_\alpha)$ is idempotent-closed for each $0 \neq \alpha \in Y_{0,n}$. Take two idempotents $f_\alpha, h_\alpha \in End(S_\alpha)$. Using Lemma 3.1.2, for every $x_\xi \in S, \xi \in Y$, take the identity map $s \in End(Y)$ and take $f, h \in End(S)$ as follows

$$f(x_\xi) := \begin{cases} f_\alpha(x_\alpha) & \text{if } \xi = \alpha, \\ e_\xi & \text{if } \xi \neq \alpha, \end{cases}$$

and

$$h(x_\xi) := \begin{cases} h_\alpha(x_\alpha) & \text{if } \xi = \alpha, \\ e_\xi & \text{if } \xi \neq \alpha. \end{cases}$$

Then f, h are idempotents. By hypothesis fh is idempotent. Then

$$\begin{aligned} f_\alpha h_\alpha f_\alpha h_\alpha(x_\alpha) &= fhfh(x_\alpha) \\ &= fh(x_\alpha) \text{ (since } fh \text{ is idempotent)} \\ &= f_\alpha h_\alpha(x_\alpha). \end{aligned}$$

Thus $f_\alpha h_\alpha$ is idempotent, and therefore $End(S_\alpha)$ is idempotent-closed for each $\alpha \in Y_{0,n}$. □

The converse is also true, which is shown below.

Lemma 3.3.2. *Let $S = [Y_{0,n}; S_\alpha, e_\alpha, c_{\alpha,e_\beta}]$ be a non-trivial strong semilattice of left simple semigroups. If the monoid $End(S_\xi)$ is idempotent-closed for all $\xi \in Y_{0,n}$, then monoid $End(S)$ is idempotent-closed.*

Proof. Take two idempotents $f, h \in End(S)$. We have $\underline{f}, \underline{h} \in End(Y_{0,n})$ are also idempotents. We now consider $\underline{f}$, $\underline{h}$.

Case 1. $\underline{f}$ and $\underline{\text{h}}$ are constant maps.

If $\underline{f}(\xi) = 0$ and $\underline{h}(\xi) = 0$ for every $\xi \in Y_{0,n}$, Then $f_0, h_0 \in End(S_0)$ and $f(S_\alpha) = h(S_\alpha) = \{f_0(e_0)\}$ for every $0 \neq \alpha \in Y_{0,n}$. This implies that $f_0h_0 \in End(S_0)$ is idempotent. Thus

$$fhfh(x_0) = f_0h_0f_0h_0(x_0) = f_0h_0(x_0) = fh(x_0)$$

and

$$fhfh(x_\alpha) = e_0 = fh(x_\alpha)$$

for every $0 \neq \alpha \in Y_{0,n}$. Therefore fh is idempotent.

If $\underline{f}(\xi) = \alpha$ for some $0 \neq \alpha \in Y_{0,n}$, then f must be a constant map, so that $fh = f$ is idempotent, and therefore fh is idempotent.

Case 2. $\underline{f}$ and $\underline{h}$ are not constant. We have in this case

$$\underline{f}\underline{h}(\alpha) := \begin{cases} \alpha & \text{if } \underline{f}(\alpha) = \underline{h}(\alpha) = \alpha, \\ 0 & \text{if } \underline{f}(\alpha) \neq \underline{h}(\alpha), \end{cases}$$

for each $\alpha \in Y_{0,n}$. In the first case we have

$$fhfh(x_\alpha) = f_\alpha h_\alpha f_\alpha h_\alpha(x_\alpha) = f_\alpha h_\alpha(x_\alpha) = fh(x_\alpha)$$

and the second case we have

$$\begin{aligned} fhfh(x_\alpha) &= fhfh_0(\varphi_{\alpha,0}(x_\alpha)) \\ &= f_0h_0f_0h_0(\varphi_{\alpha,0}(x_\alpha)) \\ &= f_0h_0(\varphi_{\alpha,0}(x_\alpha)) \\ &= fh(x_\alpha). \end{aligned}$$

Thus fh is idempotent, and therefore $End(S_\xi)$ is idempotent-closed. □

In the next theorem, we get directly from Lemmas 3.3.1 and 3.3.2.

Theorem 3.3.1. *Let $S = [Y; S_\alpha, e_\alpha, c_{\alpha,e_\beta}]$ be a non-trivial strong semilattice of left simple semigroups. Then the monoid End(S) is idempotent-closed if and only if $Y = Y_{0,n}$ and the monoid End(S_ξ) is idempotent-closed for every $\xi \in Y_{0,n}$.*

If all defining homomorphisms are isomorphisms, we have:

Theorem 3.3.2. *Let $S = [Y; T_\alpha, e_\alpha, \varphi_{\alpha,\beta}]$ be a non-trivial strong semilattice of left simple semigroups T_α with isomorphisms $\varphi_{\alpha,\beta}$. Then the monoid End(S) is idempotent-closed if and only if $Y = Y_{0,n}$ and the monoid End(T) is idempotent-closed.*

Proof. Necessity follows from Lemma 3.2.3 and *End*(Y) is idempotent-closed if and only if $Y = Y_{0,n}$ by Proposition 2.2.1.

We verify that *End*(T) is idempotent-closed. Take two idempotents $g, k \in End(T)$. Using Construction 3.1.2, take $f, h \in End(S)$ by

$$f(x_\alpha) := (g(x))_\alpha$$

and

$$h(x_\alpha) := (k(x))_\alpha$$

for every $x_\alpha \in S$, $\alpha \in Y_{0,n}$. Then f, h are idempotents. By hypothesis, fh is idempotent. Since $(gk(x))_\alpha = gk(x_\alpha)$ for $x = x_\alpha \in G_\alpha$, we have $(gkgk(x))_\alpha = gkg((k(x))_\alpha) = fhf(h(x_\alpha)) = fh(x_\alpha) = (gk(x))_\alpha$. Therefore gk is idempotent. Hence *End*(T) is idempotent-closed.

Sufficiency. Take two idempotents $f, h \in End(S)$. Then $\underline{f}, \underline{h} \in End(Y)$ which are idempotents, of course $End(Y_{0,n})$ is idempotent-closed implies that $\underline{fhfh} = \underline{fh}$, and $f|_{T_\alpha} = g$, $h|_{T_\alpha} = k$ for some idempotents $g, k \in End(T)$. Then $gk \in End(T)$ is idempotent. Thus

$$fhfh(x_\xi) = (gkgk(x))_{\underline{fhfh}(\xi)} = (gk(x))_{\underline{fh}(\xi))} = fh(x_\xi),$$

and then fh is idempotent.

Therefore *End*(S) is idempotent-closed. □

Problem 3.3.1. Investigate left simple semigroups with idempotent-closed endomorphism monoid

3.4 Orthodox monoids

In this section we consider the strong semilattices of left simple semigroups whose endomorphism monoids are orthodox.

In the next theorem we get directly from Theorems 3.2.1, and 3.2.2, 3.3.1.

Theorem 3.4.1. *Let $S = [Y; S_\alpha, e_\alpha, c_{\alpha,e_\beta}]$ be a non-trivial strong semilattice of left simple semigroups. If the monoid End(S) is orthodox then the following conditions hold*

1) $Y = Y_{0,n}$,

2) the set $Hom(S_0, S_\alpha)$ consists of constant mappings for all $\alpha \in Y_{0,n}$, $\alpha \neq 0$,

3) the monoid $End(S_\xi)$ is idempotent-closed for all $\xi \in Y_{0,n}$, and

4) the set $Hom(S_\alpha, S_\beta)$ is hom-regular for every $\alpha, \beta \in Y_{0,n}$.

The converse is also true if we add the condition that S_0 contains only one idempotent which is equivalent to S_0 is a group since we consider only the finite case.

Theorem 3.4.2. *Let $S = [Y; S_\alpha, e_\alpha, c_{\alpha,e_\beta}]$ be a non-trivial strong semilattice of left simple semigroups. If the following conditions hold*

1) $Y = Y_{0,n}$,

2) the set $Hom(S_0, S_\alpha)$ consists of constant mappings for all $\alpha \in Y_{0,n}$, $\alpha \neq 0$,

3) the monoid $End(S_\xi)$ is idempotent-closed for all $\xi \in Y_{0,n}$,

4) the set $Hom(S_\alpha, S_\beta)$ is hom-regular for every $\alpha, \beta \in Y_{0,n}$, and

5) S_0 contains one idempotent e_0,

then the monoid End(S) is orthodox.

If all defining homomorphisms are isomorphisms, we have:

Theorem 3.4.3. *Let $S = [Y; T_\alpha, e_\alpha, \varphi_{\alpha,\beta}]$ be a non-trivial strong semilattice of left simple semigroups T_α with isomorphisms $\varphi_{\alpha,\beta}$. Then the monoid End(S) is orthodox if and only $Y = Y_{0,n}$ and the monoid End(T) is orthodox.*

Problem 3.4.1. Investigate left simple semigroups S with orthodox endomorphism monoid

3.5 Left inverse monoids

In this section we consider the strong semilattices of left simple semigroups whose endomorphism semigroups are left inverse.

Lemma 3.5.1. *Let* $S = [Y; S_\alpha, e_\alpha, c_{\alpha,e_\beta}]$ *be a non-trivial strong semilattice of left simple semigroups. If the monoid* $End(S)$ *is left inverse, then* $Y = Y_{0,n}$ *and the monoid* $End(S_\xi)$ *is left inverse for every* $\xi \in Y_{0,n}$.

Proof. Necessity follows from Lemma 3.2.3 and the monoid $End(Y)$ is left inverse if and only if $Y = Y_{0,n}$ by Proposition 2.2.1.

We first show that $End(S_0)$ is left inverse. Take two idempotents $f_0, h_0 \in End(T_0)$. Using Construction 3.3.1, for every $x_\xi \in S,\ \xi \in Y_{0,n}$, take $f, h \in End(S)$ as follows

$$f(x_\xi) := \begin{cases} f_0(x_0) & \text{if } \xi = 0, \\ f_0(\varphi_{\xi,0}(x_\xi)) & \text{if } \xi \neq 0, \end{cases}$$

and

$$h(x_\xi) := \begin{cases} h_0(x_0) & \text{if } \xi = 0, \\ h_0(\varphi_{\xi,0}(x_\xi)) & \text{if } \xi \neq 0. \end{cases}$$

Thus f, h are idempotents. By hypothesis $fhf(x_\xi) = fh(x_\xi)$ for all $x_\xi \in S,\ \xi \in Y_{0,n}$. This implies $f_0h_0f_0(x_0) = fhf(x_0) = fh(x_0) = f_0h_0(x_0)$, and therefore $End(S_0)$ is left inverse.

For $0 \neq \alpha \in Y_{0,n}$. We show that $End(S_\alpha)$ is left inverse. Take two idempotents $f_\alpha, h_\alpha \in End(S_\alpha)$. Using Lemma 3.1.2, for every $x_\xi \in S,\ \xi \in Y_{0,n}$, take $f, h \in End(S)$ as follows

$$f(x_\xi) := \begin{cases} f_\alpha(x_\alpha) & \text{if } \xi = \alpha, \\ e_\xi & \text{if } \xi \neq \alpha, \end{cases}$$

and

$$h(x_\xi) := \begin{cases} h_\alpha(x_\alpha) & \text{if } \xi = \alpha, \\ e_\xi & \text{if } \xi \neq \alpha. \end{cases}$$

Thus f, h are idempotents. By hypothesis $fhf(x_\xi) = fh(x_\xi)$ for all $x_\xi \in S,\ \xi \in Y_{0,n}$. This implies

$$f_\alpha h_\alpha f_\alpha(x_\alpha) = fhf(x_\alpha) = fh(x_\alpha) = f_\alpha h_\alpha(x_\alpha),$$

and therefore $End(S_\alpha)$ is left inverse. □

The converse is also true.

Lemma 3.5.2. *Let* $Y = Y_{0,n}$ *and* $S = [Y_{0,n}; S_\alpha, e_\alpha, c_{\alpha,e_\beta}]$ *be a non-trivial strong semilattice of left inverse semigroups. If the monoid* $End(S_\xi)$ *is left inverse for each* $\xi \in Y_{0,n}$*, then the monoid* $End(S)$ *is left inverse.*

Proof. Take two idempotents $f, h \in End(S)$. We have $\underline{f}, \underline{h} \in End(Y_{0,n})$ are also idempotents. We now consider $\underline{f}$, $\underline{h}$.

Case 1. $\underline{f}$ and $\underline{\mathrm{h}}$ are constant maps.

If $\underline{f}(\xi) = 0$ and $\underline{h}(\xi) = 0$ for every $\xi \in Y_{0,n}$, Then $f_0, h_0 \in End(S_0)$ and $f(S_\alpha) = h(S_\alpha) = \{f_0(e_0)\}$ for every $0 \neq \alpha \in Y_{0,n}$. Thus

$$fhf(x_0) = f_0h_0f_0(x_0) = f_0h_0(x_0) = fh(x_0)$$

and

$$fhf(x_\alpha) = e_0 = fh(x_\alpha)$$

for every $0 \neq \alpha \in Y_{0,n}$. Therefore $fhf = fh$.

If $\underline{f}(\xi) = \alpha$ for some $0 \neq \alpha \in Y_{0,n}$, then f must be a constant map, so that $fhf = fh$.

Case 2. $\underline{f}$ and $\underline{h}$ are not constant. We have in this case

$$\underline{f}\underline{h}(\alpha) := \begin{cases} \alpha & \text{if } \underline{f}(\alpha) = \underline{h}(\alpha) = \alpha, \\ 0 & \text{if } \underline{f}(\alpha) \neq \underline{h}(\alpha), \end{cases}$$

for each $\alpha \in Y_{0,n}$. In the first case we have

$$fhf(x_\alpha) = f_\alpha h_\alpha f_\alpha(x_\alpha) = f_\alpha h_\alpha(x_\alpha) = fh(x_\alpha)$$

and the second case we have

$$\begin{aligned} fhf(x_\alpha) &= fhf_0(\varphi_{\alpha,0}(x_\alpha)) \\ &= f_0h_0f_0(\varphi_{\alpha,0}(x_\alpha)) \\ &= f_0h_0(\varphi_{\alpha,0}(x_\alpha)) \\ &= fh(x_\alpha). \end{aligned}$$

Therefore $End(S)$ is left inverse. □

The following theorem follows from Lemmas 3.5.1 and 3.5.2.

Theorem 3.5.1. *Let $S = [Y; S_\alpha, e_\alpha, c_{\alpha,e_\beta}]$ be a non-trivial strong semilattice of left simple semigroups. Then the monoid End(S) is left inverse if and only if $Y = Y_{0,n}$ and the monoid $End(S_\xi)$ is left inverse for every $\xi \in Y_{0,n}$.*

If all defining homomorphisms are isomorphisms we have.

Theorem 3.5.2. *Let $S = [Y; T_\alpha, e_\alpha, \varphi_{\alpha,\beta}]$ be a non-trivial strong semilattice of left simple semigroups T_α with isomorphisms $\varphi_{\alpha,\beta}$. Then the monoid End(S) is left inverse if and only if $Y = Y_{0,n}$ and the monoid End(T) is left inverse.*

Proof. Necessity follows from Lemma 3.2.3 and the monoid $End(Y)$ is left inverse if and only if $Y = Y_{0,n}$ by Proposition 2.2.1.

We verify that $End(T)$ is left inverse. Take two idempotents $g, k \in End(T)$. Using Construction 3.1.2, take $f, h \in End(S)$ as follows

$$f(x_\alpha) := (g(x))_\alpha$$

and

$$h(x_\alpha) := (k(x))_\alpha$$

for every $x_\alpha \in S$, $\alpha \in Y_{0,n}$. Then f, h are idempotents. Then $(gkg(x))_\alpha = fh(f(x_\alpha)) = fh(x_\alpha) = (gk(x))_\alpha$. Therefore gk is idempotent. Hence $End(G)$ is left inverse.

Sufficiency. Take two idempotents $f, h \in End(S)$. Then $\underline{f}, \underline{h} \in End(Y)$ are idempotents and $\underline{fhf} = \underline{fh}$. In fact, $f_\xi(x) = f_\alpha(x) = g \in End(T)$ and $h_\xi(x) = h_\alpha(x) = k \in End(T)$. This implies

$$fhf(x_\xi) = (gkg(x))_{\underline{fhf}(\xi)} = (gk(x))_{\underline{fh}(\xi)} = fh(x_\xi),$$

and therefore $fhf = fh$. Hence the monoid $End(S)$ is left inverse. □

Problem 3.5.1. Investigate left simple semigroups with left inverse endomorphism monoids

3.6 Completely regular monoids

In this section we consider strong semilattices of left simple semigroups whose endomorphism semigroups are completely regular.

Theorem 3.6.1. *Let $S = [Y; S_\alpha, e_\alpha, c_{\alpha,e_\beta}]$ be a non-trivial strong semilattice of left simple semigroups with $\nu \wedge Y$. If the monoid End(S) is completely regular, then the following assertions hold*

1) $|Y| = 2$,

2) the set $Hom(S_\nu, S_\alpha)$ consists of constant mappings for all $\alpha \in Y$, $\nu < \alpha$, and

3) the monoid $End(S_\xi)$ is completely regular for every $\xi \in Y$.

Proof. 1) According to Lemma 3.2.3, the monoid $End(Y)$ is completely regular follows and the monoid $End(Y)$ is completely regular if and only if $|Y| \leq 2$ by Proposition 2.2.1.

2) By Lemma 3.2.4.

3) Assume $Y = \{\nu, \mu\}, \nu < \mu$.

First, we verify that $End(S_\nu)$ is completely regular. Take $f_\nu \in End(S_\nu)$. Using Construction 3.3.1, for every $x_\xi \in S$, $\xi \in Y$, take $f \in End(S)$ as follows

$$f(x_\xi) := \begin{cases} f_\nu(x_\nu) & \text{if } \xi = \nu, \\ f_\nu(\varphi_{\mu,\nu}(x_\mu)) & \text{if } \xi = \mu. \end{cases}$$

By hypothesis there exists $f' \in End(S)$ such that $ff'f = f$ and $ff' = f'f$. Thus $f_\nu f'_\nu f_\nu(x_\nu) = ff'f(x_\nu) = f(x_\nu) = f_\nu(x_\nu)$ and $f'_\nu f_\nu(x_\nu) = f_\nu f'_\nu(x_\nu)$ and therefore f_ν is completely regular. Hence $End(S_\nu)$ is completely regular.

We show that $End(S_\mu)$ is completely regular.

Take $f_\mu \in End(S_\mu)$. Using Lemma 3.1.2, for every $x_\xi \in S$, $\xi \in Y$, take $f \in End(S)$ as follows

$$f(x_\xi) := \begin{cases} f_\mu(x_\mu) & \text{if } \xi = \mu, \\ e_\nu & \text{if } \xi = \nu. \end{cases}$$

By hypothesis there exists $f' \in End(S)$ such that $ff'f = f$ and $ff' = f'f$. Thus $f_\mu f'_\mu f_\mu(x_\mu) = ff'f(x_\mu) = f(x_\mu) = f_\mu(x_\mu)$ and $f'_\mu f_\mu(x_\mu) = f_\mu f'_\mu(x_\mu)$ and therefore f_μ is completely regular. Hence $End(S_\mu)$ is completely regular. □

The following theorem shows the converse.

Theorem 3.6.2. *Let $S = [Y; S_\alpha, e_\alpha, c_{\alpha,e_\beta}]$ be a non-trivial strong semilattice of left simple semigroups and $\nu = \wedge Y$. If the following conditions hold*

1) $|Y| = 2$,

2) the set $Hom(S_\nu, S_\alpha)$ consists of constant mappings for all $\nu < \alpha \in Y$,

3) the monoid $End(S_\xi)$ is completely regular for every $\xi \in Y$, and

4) S_ν contains one idempotent e_ν,
then the monoid $End(S)$ is completely regular.

Proof. Assume $Y = \{\nu, \mu\}, \nu < \mu$.

Take $f \in End(S)$. Then $\underline{f} \in End(Y)$ such that $\underline{f}\underline{f}'\underline{f} = \underline{f}$ and $\underline{f}'\underline{f} = \underline{f}\underline{f}'$.

Case 1. $\underline{f}(\nu) = \underline{f}(\mu) = \nu$. Then $\nu = \underline{f}(\underline{f}'(\nu)) = \underline{f}'(\underline{f}(\nu)) = \underline{f}'(\nu)$. We have $f(S_\mu) = \{f_\nu(e_\nu)\}$ and $f_\nu \in End(S_\nu)$. By hypothesis, there exists $f'_\nu \in End(S_\nu)$ such that $f_\nu f'_\nu f_\nu = f_\nu$ and $f'_\nu f_\nu = f_\nu f'_\nu$ by Lemma 3.2.1 1). Using Construction 3.3.1, for every $x_\xi \in S,\ \xi \in Y_{0,n}$, take $f' \in End(S)$ as follows

$$f'(x_\xi) := \begin{cases} f'_\nu(x_\nu) & \text{if } \xi = \nu, \\ f'_\nu(\varphi_{\mu,\nu}(x_\mu)) & \text{if } \xi = \mu. \end{cases}$$

So that $ff'f = f$ and $ff' = f'f$. Therefore f is completely regular.

Case 2. $\underline{f}(\nu) = \underline{f}(\mu) = \mu$. Then $\mu = \underline{f}(\underline{f}'(\mu)) = \underline{f}'(\underline{f}(\nu)) = \underline{f}'(\mu)$. In this case $f(S_\mu) = \{f_\nu(e_\nu)\}$. By 2) $f_\nu \in Hom(S_\nu, S_\mu)$ is constant, so that f is constant, and of course f is completely regular.

Case 3. $\underline{f}(\nu) = \nu$ and $\underline{f}(\mu) = \mu$. By Lemma 3.1.3 $f_\nu(e_\nu) = f_\nu\varphi_{\mu,\nu}(x_\mu) = \varphi_{\mu,\nu}f_\mu(x_\mu) = e_\nu$, so take $f'_\nu \in End(S_\nu)$ with $f'_\nu(e_\nu) = e_\nu$ since S_ν has only one idempotent by 4). Thus there exist $f'_\nu \in End(S_\nu)$ and $f'_\mu \in End(S_\mu)$ such that $f_\nu f'_\nu f_\nu = f_\nu$ and $f'_\nu f_\nu = f_\nu f'_\nu$ and $f_\mu f'_\mu f_\mu = f_\mu$ and $f'_\mu f_\mu = f_\mu f'_\mu$. Using Lemma 3.1.2, for every $x_\xi \in S,\ \xi \in Y_{0,n}$, take $f' \in End(S)$ as follows

$$f'(x_\xi) := \begin{cases} f'_\nu(x_\nu) & \text{if } \xi = \nu, \\ f'_\mu(x_\mu) & \text{if } \xi = \mu. \end{cases}$$

Thus $ff'f = f$ and $ff' = f'f$. Therefore f is completely regular. Hence $End(S)$ is completely regular. □

If all defining homomorphisms are isomorphisms we have.

Theorem 3.6.3. *Let $S = [Y; T_\alpha, e_\alpha, \varphi_{\alpha,\beta}]$ be a non-trivial strong semilattice of semigroups T_α with isomorphisms $\varphi_{\alpha,\beta}$ and $\nu = \wedge Y$. Then the monoid $End(S)$ is completely regular if and only if $|Y| = 2$ and the monoid $End(T)$ is completely regular.*

Proof. Necessity follows from Lemma 3.2.3 and the monoid $End(Y)$ is completely regular if and only if $|Y| \le 2$ by Proposition 2.2.1.

Assume $Y = \{\nu, \mu\}, \nu < \mu$.

We show that $End(T)$ is completely regular. Take $g \in End(T)$. Using Construction 3.1.2, take $f \in End(S)$ as follows

$$f(x_\alpha) := (g(x))_\alpha$$

for every $x_\alpha \in S,\ \alpha \in Y$. By hypothesis $f' \in End(S)$ exists such that $ff'f = f$ and $ff' = f'f$. It is clear that g is completely regular since $f_\nu = f_\mu = g$. Hence $End(T)$ is completely regular.

Sufficiency. Assume $Y = \{\nu, \mu\}, \nu < \mu$. Take $f \in End(S)$. Then $\underline{f} \in End(Y)$. By hypothesis there exists $s \in End(Y)$ such that $\underline{f}s\underline{f} = \underline{f}$ and $\underline{f}s = s\underline{f}$. We have $f_\mu(x) = f_\nu(x) = g(x)$ where $g \in End(T)$ and $End(T)$ is completely regular, there exists $g' \in End(T)$ such that $gg'g = g$ and $gg' = g'g$. Using Construction 3.1.2, take $f' \in End(S)$ as follows

$$f'(x_\alpha) := (g'(x))_{s(\alpha)}$$

for every $x_\alpha \in S,\ \alpha \in Y$. Then $ff'(f(x_\alpha) = ff'((g(x))_\alpha) = (gg'g(x))_{\underline{f}s\underline{f}(\alpha)} = (g(x))_{\underline{f}(\alpha)}) = f(x_\alpha)$ and $f'f(x_\alpha) = (g'g(x))_{s\underline{f}(\alpha)} = (gg'(x))_{\underline{f}s(\alpha)} = ff'(x_\alpha)$ for $\alpha \in Y$. Then f is completely regular, and therefore $End(S)$ is completely regular. □

Problem 3.6.1. Investigate left simple semigroups with completely regular endomorphism monoid

3.7 Idempotent monoids

In this section we consider the strong semilattices of left simple semigroups whose endomorphism monoids are idempotent.

Lemma 3.7.1. *Let $S = [Y; S_\alpha, e_\alpha, c_{\alpha,e_\beta}]$ be a non-trivial strong semilattice of left simple semigroups. If the monoid $End(S)$ is idempotent, then the following hold*

1) $|Y| = 2$,

2) the set $Hom(S_\nu, S_\alpha)$ consists of constant mappings for all $\nu < \alpha \in Y$, and

3) the monoid $End(S_\xi)$ is idempotent for every $\xi \in Y$.

Proof. 1) According to Lemma 3.2.3, the monoid $End(Y)$ is idempotent and the monoid $End(Y)$ is idempotent if and only if $|Y| \leq 2$ by Proposition 2.2.1. Assume $Y = \{\nu, \mu\}, \nu < \mu$.

2) Take $f_\nu \in Hom(S_\nu, S_\mu)$. Using Lemma 3.1.2, for every $x_\xi \in S,\ \xi \in Y$, take $f \in End(S)$ as follows

$$f(x_\xi) := \begin{cases} f_\nu(x_\nu) \in S_\mu & \text{if } \xi = \nu, \\ f_\nu(e_\nu) & \text{if } \xi = \mu. \end{cases}$$

By hypothesis f is idempotent. Thus $ff(x_\nu) = f(x_\nu) = f_\nu(x_\nu) \in S_\mu$ and $f_\nu(x_\nu)$ must be equal to $f_\nu(e_\nu)$. Thus f_ν is a constant map.

3) First, we verify that $End(S_\nu)$ is idempotent. Take $f_\nu \in End(S_\nu)$. Using Construction 3.3.1, for every $x_\xi \in S,\ \xi \in Y$, take $f \in End(S)$ follows

$$f(x_\xi) := \begin{cases} f_\nu(x_\nu) & \text{if } \xi = \nu, \\ f_\nu(\varphi_{\mu,\nu}(x_\mu)) & \text{if } \xi = \mu. \end{cases}$$

By hypothesis f is idempotent. Thus

$$f_\nu f_\nu(x_\nu) = ff(x_\nu) = f(x_\nu) = f_\nu(x_\nu).$$

This implies that f_ν is idempotent. So that $End(S_\nu)$ is idempotent.

We verify now that $End(S_\mu)$ is idempotent. Take $f_\mu \in End(S_\mu)$. Using Lemma 3.1.2, for every $x_\xi \in S,\ \xi \in Y$, take $f \in End(S)$ as follows

$$f(x_\xi) := \begin{cases} f_\mu(x_\mu) & \text{if } \xi = \mu, \\ e_\nu & \text{if } \xi = \nu. \end{cases}$$

By hypothesis f is idempotent. Thus

$$f_\mu f_\mu(x_\mu) = ff(x_\mu) = f(x_\mu) = f_\mu(x_\mu).$$

This implies that f_μ is idempotent. So that $End(S_\mu)$ is idempotent. □

The converse is also true.

Lemma 3.7.2. *Let $S = [Y; S_\alpha, e_\alpha, c_{\alpha,e_\beta}]$ be a non-trivial strong semilattice of left simple semigroups. If the following conditions hold*

1) $|Y| = 2$,

2) the set $Hom(S_\nu, S_\alpha)$ consists of constant mappings for all $\nu < \alpha \in Y$, and

3) the monoid $End(S_\xi)$ is idempotent for each $\xi \in Y$,

then the monoid $End(S)$ is idempotent.

Proof. Assume $Y = \{\nu, \mu\}, \nu < \mu$. Take $f \in End(S)$. Then $\underline{f} \in End(Y)$ which is idempotent because $|Y| \leq 2$. We now consider three cases.

Case 1. $\underline{f}(\nu) = \underline{f}(\mu) = \nu$. We have $f(S_\mu) = \{f_\nu(e_\nu)\}$. Then $ff(x_\nu) = f_\nu f_\nu(x_\nu) = f_\nu(x_\nu) = f(x_\nu)$ where $f_\nu \in End(S_\nu)$ and $ff(x_\mu) = f(e_\nu) = e_\nu = f(x_\mu)$. Thus f is idempotent.

Case 2. $\underline{f}(\nu) = \underline{f}(\mu) = \mu$. By 2) $f_\nu \in Hom(S_\nu, S_\mu)$ is constant and $f(S_\mu) = \{f_\nu(e_\nu)\}$. This implies f is constant and of course f is idempotent.

Case 3. $\underline{f}(\nu) = \nu$ and $\underline{f}(\mu) = \mu$. We have $f_\nu \in End(S_\nu)$ and $f_\mu \in End(S_\mu)$, and $End(S_\nu)$ and $End(S_\mu)$ are idempotents, so that $ff(x_\nu) = f_\nu f_\nu(x_\nu) = f_\nu(x_\nu) = f(x_\nu)$ and $ff(x_\mu) = f_\mu f_\mu(x_\mu) = f_\mu(x_\mu) = f(x_\mu)$. Therefore f is idempotent. Hence $End(S)$ is idempotent. □

The following theorem follows from Lemmas 3.7.1 and 3.7.2.

Theorem 3.7.1. *Let $S = [Y; S_\alpha, e_\alpha, c_{\alpha,e_\beta}]$ be a non-trivial strong semilattice of left simple semigroups and $\nu = \wedge Y$. Then the monoid End(S) is idempotent if and only if the following conditions hold*

1) $|Y| = 2$, and

2) the set $Hom(S_\nu, S_\alpha)$ consists of constant mappings for all $\nu < \alpha \in Y$, and

3) the monoid $End(S_\xi)$ is idempotent for each $\xi \in Y$.

If all defining homomorphisms are isomorphisms we have.

Theorem 3.7.2. *Let $S = [Y; T_\alpha, e_\alpha, \varphi_{\alpha,\beta}]$ be a non-trivial strong semilattice of left simple semigroups T_α with isomorphisms $\varphi_{\alpha,\beta}$ and $\nu = \wedge Y$. Then the monoid End(S) is idempotent if and only if $|Y| = 2$ and the monoid End(T) is idempotent.*

Proof. Necessity follows from Lemma 3.2.3 and the monoid *End*(*Y*) is idempotent if and only if $|Y| \leq 2$ by Proposition 2.2.1.

Assume that $Y = \{\nu, \mu\}$ with $\nu < \mu$.

We show that *End*(*T*) is idempotent. Take $g \in End(T)$. Using Construction 3.1.2, take $f \in End(S)$ as follows

$$f(x_\xi) := (g(x))_\alpha$$

for every $x_\alpha \in S$, $\alpha \in Y$. By hypothesis f is idempotent. Then

$$\begin{aligned}(gg(x))_\alpha &= f(f(x_\alpha)) \\ &= f(x_\alpha) \text{ (since } f \text{ is idempotent)} \\ &= g((x))_\alpha.\end{aligned}$$

Thus g is idempotent, and therefore $End(T)$ is idempotent.

Sufficiency. Assume $Y = \{\nu, \mu\}, \nu < \mu$. Take $f \in End(S)$. Then $\underline{f} \in End(Y)$ is idempotent, implies that $\underline{f}\ \underline{f} = \underline{f}$ and $f|_{T_\alpha} = g$ for some idempotent $g \in End(T)$. Then

$$ff(x_\alpha) = (gg(x))_{\underline{f}\ \underline{f}(\alpha)} = (g(x))_{\underline{f}(\alpha)} = f(x_\alpha).$$

Thus f is idempotent, and therefore $End(S)$ is idempotent. □

Problem 3.7.1. Investigate left simple semigroups with idempotent endomorphism monoids

Chapter 4

Endomorphisms of Clifford semigroups with constant or bijective defining homomorphisms

In this chapter, we study our usual properties of the endomorphism monoids of Clifford semigroups such as regular, idempotent-closed, orthodox, left inverse, completely regular, and idempotent. Some results of this chapter have been in [8]. The Clifford semigroups have various equivalent definitions: as completely regular semigroups where elements of the form xx^{-1} commute; as regular semigroups whose idempotents are central; as semilattices of groups; or, in the way used most in this thesis, as strong semilattices of groups (see [14]). For the definition of a Clifford semigroup as a strong semilattice of groups is Definition 1.1.5 and Theorem 1.1.2.

We collect the results of this chapter in the Overview.

4.1 Regular endomorphisms

In this section the endomorphism monoids of Clifford semigroups with constant defining homomorphisms $\varphi_{\alpha,\beta}$ and Y as a finite chain are studied [21].

The following corollary is a consequence of Theorem 3.2.1. The Condition 3) of Theorem 3.2.1 is deduced.

Corollary 4.1.1. *Let $Y = Y_{0,n}$ and let $S = [Y; G_\alpha, e_\alpha, c_{\alpha,e_\beta}]$ be a Clifford semigroup. If the following conditions hold*

1) $|Hom(G_0, G_\alpha)| = 1$ for all $\alpha \in Y_{0,n}$ with $\alpha \neq 0$, and

2) the set $Hom(G_\alpha, G_\beta)$ is hom-regular for every $\alpha, \beta \in Y_{0,n}$

then the monoid End(S) is regular.

Proof. See Theorem 3.2.1. □

Corollary 4.1.2. *Let $S = [Y; G_\alpha, e_\alpha, c_{\alpha,e_\beta}]$ be a Clifford semigroup with $\nu = \wedge Y$. If the monoid End(S) is regular then the following conditions hold*

1) the monoid End(Y) is regular,

2) $|Hom(G_\nu, G_\alpha)| = 1$ for all $\alpha \in Y$ with $\nu < \alpha$, and

3) the set $Hom(G_\alpha, G_\beta)$ is hom-regular for every $\alpha, \beta \in Y$.

Proof. See Theorem 3.2.2. □

If all the defining homomorphisms are bijective, we have.

Corollary 4.1.3. *Let $S = [Y; G_\alpha, e_\alpha, \varphi_{\alpha,\beta}]$ be a Clifford semigroup with bijective $\varphi_{\alpha,\beta}$. Then the monoid End(S) is regular if and only if the monoids End(Y) and End(G) are regular.*

Proof. See Theorem 3.2.3. □

Problem 4.1.1. Condition 2) of Corollary 4.1.2 is easy to check, but we do not know much about the Condition 3) nor about regularity of *End*(*G*).

4.2 Idempotent-closed monoids

In this section we investigate Clifford semigroups whose endomorphism monoids are idempotent-closed.

Corollary 4.2.1. *Let $S = [Y; G_\alpha, e_\alpha, c_{\alpha,e_\beta}]$ be a Clifford semigroup. Then the monoid End(S) is idempotent-closed if and only if $Y = Y_{0,n}$ and the monoid $End(G_\xi)$ is idempotent-closed for all $\xi \in Y_{0,n}$.*

Proof. See Theorem 3.3.1. □

Corollary 4.2.2. *Let $S = [Y; G_\alpha, \varphi_{\alpha,\beta}]$ be a Clifford semigroup with bijective $\varphi_{\alpha,\beta}$. Then the monoid End(S) is idempotent-closed if and only if $Y = Y_{0,n}$ and the monoid End(G) is idempotent-closed.*

Proof. See Theorem 3.3.2. □

Example 4.2.1. For any group G, the monoid $End(G)$ which is idempotent-closed, have not been found in any literature. However, we know $(End(\mathbb{Z}_n), \circ) \cong (\mathbb{Z}_n, \cdot)$ (see [9]) and $(\mathbb{Z}_n, \cdot)$ is a commutative semigroup, this implies that it is idempotent-closed and the monoid $End(\mathbb{Z}_2 \times \mathbb{Z}_2)$ is one example which is not idempotent-closed (see also Example 1.1.2*)*.

Problem 4.2.1. Investigate a group whose endomorphism monoid is idempotent-closed.

4.3 Orthodox monoids

In this section we characterize Clifford semigroups whose endomorphism monoids are orthodox.

The following corollary follows from Corollaries 4.1.1 and 4.2.1.

Corollary 4.3.1. *Let $S = [Y; G_\alpha, e_\alpha, c_{\alpha,e_\beta}]$ be a Clifford semigroup. Then the monoid $End(S)$ is orthodox if and only if the following conditions hold:*

1) $Y = Y_{0,n}$,

2) $|Hom(G_0, G_\alpha)| = 1$ for all $\alpha \in Y_{0,n}$ with $\alpha \neq 0$,

3) the monoid $End(G_\xi)$ is idempotent-closed for all $\xi \in Y_{0,n}$, and

4) the set $Hom(G_\alpha, G_\beta)$ is hom-regular for all $\alpha, \beta \in Y_{0,n}$.

Proof. See Corollaries 4.1.1 and 4.2.1. □

Now all the defining homomorphisms are bijective, we have.

Corollary 4.3.2. *Let $S = [Y; G_\alpha, \varphi_{\alpha,\beta}]$ be a Clifford semigroup with bijective $\varphi_{\alpha,\beta}$. Then the monoid $End(S)$ is orthodox if and only if $Y = Y_{0,n}$ and the monoid $End(G)$ is orthodox.*

Proof. See Corollaries 4.1.2 and 4.2.2. □

Problem 4.3.1. Investigate a group whose endomorphism monoid is orthodox.

4.4 Left inverse monoids

In this section we study Clifford semigroups whose endomorphism monoids are left inverse.

Corollary 4.4.1. *Let $S = [Y; G_\alpha, e_\alpha, c_{\alpha,e_\beta}]$ be a Clifford semigroup. Then the monoid $End(S)$ is left inverse if and only if $Y = Y_{0,n}$ and the monoid $End(G_\xi)$ is left inverse for all $\xi \in Y_{0,n}$.*

Proof. See Theorem 3.5.1. □

Now all the defining homomorphisms are bijective, we have.

Corollary 4.4.2. *Let $S = [Y; G_\alpha, \varphi_{\alpha,\beta}]$ be a Clifford semigroup with bijective $\varphi_{\alpha,\beta}$. Then the monoid $End(S)$ is left inverse if and only if $Y = Y_{0,n}$ and the monoid $End(G)$ is left inverse.*

Proof. See Theorem 3.5.2. □

Problem 4.4.1. Investigate a group whose endomorphism monoid is left inverse.

For a group G whose endomorphism monoid is inverse, has been investigated in [10]. We collect some results also here. The proofs can be found in [10].

Definition 4.4.1. We call a group G with the property that $End(G)$ is an inverse semigroup, an *inverse group.* If $End(G) = Aut(G) \cup \{0\}$ we call G a *basic group.*

Proposition 4.4.1. *Let G be an inverse group. Let $f \in End(G)$. Then $Ker(f)$ has a unique complement.*

Proposition 4.4.2. *Let G be an inverse group. Then either G is basic or $G = H \oplus K$ where H and K are fully invariant subgroups of G, H and K are both inverse groups and $Hom(H, K) = 0$.*

Proposition 4.4.3. *Let H and K be inverse groups such that $Hom(H, K) = 0$, $Hom(K, H) = 0$. Then $H \oplus K$ is an inverse group.*

4.5 Completely regular monoids

We now consider Clifford semigroups whose endomorphism monoids are completely regular.

Corollary 4.5.1. *Let $S = [Y; G_\alpha, e_\alpha, c_{\alpha,e_\beta}]$ be a Clifford semigroup. Then the monoid End(S) is completely regular if and only if*

1) $|Y| = 2$, or $Y = \{\nu, \mu\}$ with $\nu < \mu$,

2) $|Hom(G_\nu, G_\mu)| = 1$, and

3) the monoids $End(G_\nu)$ and $End(G_\mu)$ are completely regular.

Proof. See Theorems 3.6.1 and 3.6.2. □

Now all the defining homomorphisms are bijective, we have.

Corollary 4.5.2. *Let $S = [Y; G_\alpha, \varphi_{\alpha,\beta}]$ be a Clifford semigroup with bijective $\varphi_{\alpha,\beta}$. Then the monoid End(S) is completely regular if and only if $|Y| = 2$ and the monoid End(G) is completely regular.*

Proof. See Theorem 3.6.3. □

Problem 4.5.1. Investigate a group whose endomorphism monoid is completely regular.

4.6 Idempotent monoids

We discuss now Clifford semigroups whose endomorphism monoids are idempotent.

The following is folklore.

Lemma 4.6.1. *The endomorphism monoid of a group G is idempotent if and only if $G = \mathbb{Z}_1$ or $G = \mathbb{Z}_2$*

Proof. Sufficiency. It is obvious.

Necessity. Suppose that $|G| > 2$. It can be define $f : G \to G$ as follows

$$f(x) := axa^{-1}$$

for every $x \in G$, for some $a \in G$ such that a^{-1} is the inverse of a. It is easily to check that f is an isomorphism of G and f is not the identity map. □

Corollary 4.6.1. *Let $S = [Y; G_\alpha, e_\alpha, c_{\alpha,e_\beta}]$ be a Clifford semigroup. Then the monoid End(S) is idempotent if and only if the following hold*

1) $|Y| = 2$, i.e., $Y = \{\nu, \mu\}$, $\nu < \mu$,

2) $G_\nu, G_\mu \in \{\mathbb{Z}_1, \mathbb{Z}_2\}$, *and*

3) $|Hom(G_\nu, G_\alpha)| = 1$, *i.e.,* $G_\nu \neq G_\mu$.

Proof. See Theorem 3.7.1. □

Now all the defining homomorphisms are bijective, we have.

Corollary 4.6.2. *Let* $S = [Y; G_\alpha, \varphi_{\alpha,\beta}]$ *be a Clifford semigroup with bijective* $\varphi_{\alpha,\beta}$. *Then the monoid* $End(S)$ *is idempotent if and only if* $|Y| = 2$ *and the monoid* $End(G)$ *is idempotent.*

Proof. See Theorem 3.7.2. □

Problem 4.6.1. Investigate a group whose endomorphism monoid is idempotent.

Chapter 5

Endomorphisms of strong semilattices of left groups

In this chapter we discuss strong semilattices of left groups with constant defining homomorphisms and isomorphisms $\varphi_{\alpha,\beta}$ whose endomorphism monoids are idempotent-closed, regular, orthodox, left inverse, completely regular, and idempotent.

It is well-known that a left group S is isomorphic to a direct product of a group G and a left zero semigroup L_n for some a positive integer n. We denote the left zero semigroup by $L_n = \{l_1, l_2, ..., l_n\}$ such that $l_i l_j = l_i$ for $i, j \in \{1, 2, ..., n\}$ and written $S = L_n \times G$ as a left group. Dually we denote by $T = H \times R_m$ a right group where H is a group and $R_m = \{1, 2, 3, ..., m\}$ is a right zero semigroup such that $r_i r_j = r_j$ for $i, j \in \{1, 2, 3, ..., m\}$.

Denoted by $E(L_n \times G) = \{(l, e_G) \mid l \in L_n \text{ and } e_G \text{ is the identity of } G\}$ the set of idempotents of the left group $L_n \times G$.

In the case of constant defining homomorphisms, we denoted by (l'_β, e_β) a fixed element in $L_{n_\beta} \times G_\beta$ such that $\varphi_{\alpha,\beta}((l_\alpha, x_\alpha)) = (l'_\beta, e_\beta)$ for every $(l_\alpha, x_\alpha) \in L_{n_\alpha} \times G_\alpha$.

We collect the results in this chapter as a table in the Overview.

5.1 Regular monoids

We first provide some auxiliary results which are needed later.

Construction 5.1.1. *Let $L_n \times G$ and $L_m \times H$ be left groups. Take $g \in Hom(G, H)$ and $s \in Hom(L_n, L_m)$. Define $f : L_n \times G \to L_m \times H$ by*

$$f((l, x)) := (s(l), g(x))$$

for every $(l, x) \in L_n \times G$. *Then* $f \in Hom(L_n \times G, L_m \times H)$.

Conversely, if $f \in Hom(L_n \times G, L_m \times H)$, *then* $p_1 f i_1 \in Hom(L_n, L_m)$ *and* $p_2 f i_2 \in Hom(G, H)$ *where* $p_1 : L_m \times H \to L_m$ *is the first projection map,* $p_2 : L_m \times H \to H$ *is the second projection map, and* $i_1 : L_n \to L_n \times G$ *is the first embedding map,* $i_2 : G \to L_n \times G$ *is the second embedding map.*

Proof. It can be seen that f is well-defined. We show that f is a homomorphism. Take $(l, x), (l', y) \in L_n \times G$. Then

$$\begin{aligned} f((l,x)(l',y)) &= f((l,xy)) \\ &= (s(l), g(xy)) \\ &= (s(l)s(l'), g(x)g(y)) \\ &= (s(l), g(x))(s(l'), g(y)) \\ &= f((l,x))f((l',y)). \end{aligned}$$

Thus $f \in Hom(L_n \times G, L_m \times H)$.

Conversely, let $f \in Hom(L_n \times G, L_m \times H)$. Then $p_2 f i_2 \in Hom(G, H)$ and $p_1 f i_1 \in Hom(L_n, L_m)$. □

Corollary 5.1.1. *Let* $L_n \times G$ *and* $L_m \times H$ *be left groups. Then* $Hom(L_n \times G, L_m \times H) \cong Hom(L_n, L_m) \times Hom(G, H)$.

We remark that the above Construction is also true for the right groups, so we have the next corollary.

Construction 5.1.2. *Let* $G \times R_n$ *and* $H \times R_m$ *be left groups. Take* $g \in Hom(G, H)$ *and* $s \in Hom(R_n, R_m)$. *Define* $f : G \times R_n \to H \times R_m$ *by*

$$f((x, r)) := (g(x), s(r))$$

for every $(x, r) \in G \times R_n$. *Then* $f \in Hom(G \times R_n, H \times R_m)$.

Conversely, if $f \in Hom(G \times R_n, H \times R_m)$, *then* $p_1 f i_1 \in Hom(G, H)$ *and* $p_2 f i_2 \in Hom(R_n, R_m)$ *where* $p_1 : H \times R_m \to G$ *is the first projection map,* $p_2 : H \times R_m \to R_m$ *is the second projection map and* $i_1 : G \to G \times R_n$ *is the first embedding map,* $i_2 : R_n \to G \times R_n$ *is the second embedding map.*

Corollary 5.1.2. *Let* $G \times R_n$ *and* $H \times R_m$ *be right groups. Then* $Hom(G \times R_n, H \times R_m) \cong Hom(G, H) \times Hom(R_n, R_m)$.

From now on we prove only the case of non-trivial left groups $L_n \times G$, that is $n \geq 2$, but we will get the results for right groups as well.

Lemma 5.1.1. *Take any left groups $L_n \times G$ and $L_m \times H$. If the set $Hom(L_n \times G, L_m \times H)$ consists of constant mappings, then $|Hom(G, H)| = 1$.*

Proof. Let $g \in Hom(G, H)$. Using Construction 5.1.1, take $f \in Hom(L_n \times G, L_m \times H)$ as follows

$$f((l, x)) := (l, g(x))$$

for every $(l, x) \in L_n \times G$. By hypothesis, $Hom(L_n \times G, L_m \times H)$ consists of constant mappings. This means $(l', g(x)) = f((l, x)) = (m, e_H)$ for some idempotent $(m, e_H) \in E(L_m \times H)$. This implies $l' = m,\ g(x) = e_H$ for every $x \in G$ where e_H is the identity in H. Hence $|Hom(G, H)| = 1$. □

By taking $G = H,\ n = m$ in Corollary 5.1.2, we have the following.

Corollary 5.1.3. *Let $L_n \times G$ be a left group. Then $End(L_n \times G) \cong End(L_n) \times End(G)$.*

Lemma 5.1.2. *Take a left group $L_n \times G$. Then the monoid $End(L_n \times G)$ is regular (idempotent-closed, orthodox, left inverse, completely regular, and idempotent) if and only if the monoids $End(G)$ and $End(L_n)$ are regular (idempotent-closed, orthodox, left inverse, completely regular, and idempotent).*

Proof. This is clear since $End(L_n \times G) \cong End(L_n) \times End(G)$ by Corollary 5.1.3. □

We repeat the property of the monoid $End(L_n)$ which is equivalent to Corollary 2.2.3.

Lemma 5.1.3. *Take a left zero semigroup L_n. Then the monoid $End(L_n)$ is*

$$\begin{array}{ll} 1) & \textit{always regular,} \\ 2) & \textit{completely regular} \\ 3) & \textit{idempotent-closed} \\ 4) & \textit{orthodox} \\ 5) & \textit{left inverse} \end{array} \left.\begin{array}{l} \\ \\ \\ \\ \end{array}\right\} \textit{iff } n = 2$$

$$\left.\begin{array}{ll} 6) & \textit{right inverse} \\ 7) & \textit{inverse} \\ 8) & \textit{a group} \\ 9) & \textit{commutative} \\ 10) & \textit{idempotent} \end{array}\right\} \textit{iff } n = 1.$$

Corollary 5.1.4. *Take a non trivial left group $L_n \times G$. Then the monoid $End(L_n \times G)$ is always regular if $End(G)$ is regular.*

The monoid $End(L_n \times G)$ has the properties 2)-5) if and only if $n = 2$ and $End(G)$ has the corresponding property.

The monoid $End(L_n \times G)$ has the properties 6)-10) if and only if $n = 1$ and $End(G)$ has the corresponding property.

Lemma 5.1.4. *Let $S = [Y; L_{n_\alpha} \times G_\alpha, \varphi_{\alpha,\beta}]$ be a non-trivial strong semilattice of left groups $L_{n_\alpha} \times G_\alpha$. If the monoid $End(S)$ is regular (idempotent-closed, orthodox, left inverse, completely regular, and idempotent), then the monoid $End(Y)$ is regular (idempotent-closed, orthodox, left inverse, completely regular, and idempotent).*

Proof. See Lemma 3.2.3. □

Lemma 5.1.5. *Let $S = [Y; L_{n_\alpha} \times G_\alpha, \varphi_{\alpha,\beta}]$ be a non-trivial strong semilattice of left groups $L_{n_\alpha} \times G_\alpha$ with constant defining homomorphisms $\varphi_{\alpha,\beta}$. If the monoid $End(S)$ is regular (completely regular, idempotent-closed, orthodox, idempotent, left inverse), then the monoid $End(L_{n_\alpha} \times G_\alpha)$ is regular (completely regular, idempotent-closed, orthodox, idempotent, left inverse).*

Proof. See Lemma 3.2.2. □

Lemma 5.1.6. *If the set $Hom(L_n \times G, L_m \times H)$ is hom-regular if and only if the sets $Hom(G, H)$ and $Hom(L_n, L_m)$ are hom-regular.*

Proof. As a consequence of Corollary 5.1.1, the set $Hom(L_n \times G, L_m \times H) \cong Hom(G, H) \times Hom(L_n, L_m)$. □

Lemma 5.1.7. *The set $Hom(L_n, L_m)$ is always hom-regular.*

Proof. Take $f \in Hom(L_n, L_m)$. Define $f' : L_m \to L_n$ as follows

$$f'(x) := \begin{cases} l' & \text{if } x \in Im(f), \text{ for some } l' \in f^{-1}\{x\} \\ l_1 & \text{if } x \notin Im(f), \ l_1 \in L_n. \end{cases}$$

for every $x \in L_m$. Then $f' \in Hom(L_m, L_n)$ such that $ff'f = f$. □

We discuss the strong semilattices of left groups whose endomorphism monoids are regular. The following corollary is a consequence of Theorem 3.2.1. The Condition 2) of Theorem 3.2.1 is deduced by Lemma 5.1.6 and the set $Hom(L_n, L_m)$ is always regular by Lemma 5.1.7. The Condition 3) of Theorem 3.2.1 is deduced since $L_{n0} \times G_0$ must contain only one idempotent, implies that $L_{n0} = L_1$.

Corollary 5.1.5. *Let $Y = Y_{0,m}$ and let $S = [Y; L_{n_\alpha} \times G_\alpha, c_{\alpha,(l_\beta,e_\beta)}]$ be a non-trivial strong semilattice of left groups $L_{n_\alpha} \times G_\alpha$. If the following conditions hold*

1) $|Hom(L_{n0} \times G_0, L_\alpha \times G_\alpha)| = 1$ for all $\alpha \in Y_{0,m}$ with $\alpha \neq 0$,

2) the set $Hom(G_\alpha, G_\beta)$ is hom-regular for every $\alpha, \beta \in Y_{0,m}$ and,

3) $|L_{n0}| = 1$,

then the monoid $End(S)$ is regular.

Proof. See Theorem 3.2.1. □

The following corollary is a consequence of Theorem 3.2.2. The Condition 3) of Theorem 3.2.2 should be *the set $Hom(L_{n_\alpha} \times G_\alpha, L_{n_\beta} \times G_\beta)$ is hom-regular for every $\alpha, \beta \in Y$*. But this condition will be deduced by Lemmas 5.1.6 and 5.1.7. We have

Corollary 5.1.6. *Let $S = [Y; L_{n_\alpha} \times G_\alpha, c_{\alpha,(l_\beta,e_\beta)}]$ be a non-trivial strong semilattice of left groups $L_{n_\alpha} \times G_\alpha$ with $\nu = \wedge Y$. If the monoid $End(S)$ is regular then the following conditions hold*

1) the monoid $End(Y)$ is regular, i.e., Y is a binary tree or Y has only one $\wedge$-reducible or $Y \in \mathbf{B} \cup \mathbf{B}^d \cup \mathbf{R}$ (see Theorem 2.1.1),

2) the set $Hom(L_{n_\nu} \times G_\nu, L_{n_\alpha} \times G_\alpha)$ consists of constant mappings for all $\nu < \alpha \in Y$, and

3) the set $Hom(G_\alpha, G_\beta)$ is hom-regular for every $\alpha, \beta \in Y$.

Proof. See Theorem 3.2.2. □

The following example is easily to see that the Condition 3) of Corollary 5.1.5 is needed.

Example 5.1.1. Let $Y = \{0, \alpha\}$ with $0 < \alpha$ and let S be a strong semilattice of left groups T_0 and T_α with constant defining homomorphism $\varphi_{\alpha,0} = c_{(l_1,0)}$ where $T_0 = L_2 \times \mathbb{Z}_1$ and $T_\alpha = L_2 \times \mathbb{Z}_2$ such that $|L_{n_0}| = 2$. Take f as follows.

$$f = \begin{pmatrix} (l_{1_0}, 0_0) & (l_{2_0}, 0_0) & (l_{1_\alpha}, 0_\alpha) & (l_{1_\alpha}, 1_\alpha) & (l_{2_\alpha}, 0_\alpha) & (l_{2_\alpha}, 1_\alpha) \\ (l_{1_\alpha}, 0_\alpha) & (l_{2_\alpha}, 0_\alpha) & (l_{1_\alpha}, 0_\alpha) & (l_{1_\alpha}, 0_\alpha) & (l_{1_\alpha}, 0_\alpha) & (l_{1_\alpha}, 0_\alpha) \end{pmatrix}$$

Then f has no an inverse element in $End(S)$. This implies that $End(S)$ is not regular.

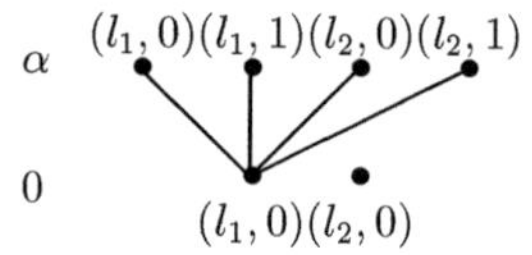

Now the defining homomorphisms are bijective:

Corollary 5.1.7. *Let $S = [Y; L_{n_\alpha} \times G_\alpha, \varphi_{\alpha,\beta}]$ be a non-trivial strong semilattice of left groups $L_{n_\alpha} \times G_\alpha$ with bijective $\varphi_{\alpha,\beta}$. Then the monoid $End(S)$ is regular if and only if the monoids $End(Y)$ and $End(L_n \times G)$ are regular.*

Proof. See Theorem 3.2.3. □

Since $End(L_n)$ is always regular and $End(L_n \times G) \cong End(L_n) \times End(G)$, we formulate Corollary 5.1.7 as follows.

Corollary 5.1.8. *Let $S = [Y; L_{n_\alpha} \times G_\alpha, \varphi_{\alpha,\beta}]$ be a non-trivial strong semilattice of left groups $L_{n_\alpha} \times G_\alpha$ with bijective $\varphi_{\alpha,\beta}$. Then the monoid $End(S)$ is regular if and only if the monoids $End(Y)$ and $End(G)$ are regular.*

5.2 Idempotent-closed monoids

In this section we consider the strong semilattices of left groups whose endomorphism monoids are idempotent-closed.

Corollary 5.2.1. *Let $S = [Y; L_{n_\alpha} \times G_\alpha, c_{\alpha,(l_\beta,e_\beta)}]$ be a non-trivial strong semilattice of left groups $L_{n_\alpha} \times G_\alpha$ with $\nu = \wedge Y$. Then the monoid $End(S)$ is idempotent-closed if and only if $Y = Y_{0,m}$, $n_\alpha = 2$ and the monoid $End(G_\alpha)$ is idempotent-closed for all $\alpha \in Y_{0,m}$.*

Proof. See Theorem 3.3.1 and Lemma 5.1.2. □

Now the defining homomorphisms are bijective.

Corollary 5.2.2. *Let $S = [Y; L_{n_\alpha} \times G_\alpha, \varphi_{\alpha,\beta}]$ be a non-trivial strong semilattice of left groups $L_{n_\alpha} \times G_\alpha$ with bijective $\varphi_{\alpha,\beta}$. Then the monoid $End(S)$ is idempotent-closed if and only if $Y = Y_{0,m}$, $n = 2$ and the monoid $End(G)$ is idempotent-closed.*

Proof. By Theorem 3.3.2 and Lemma 5.1.2. □

The following example shows positive and negative for Corollary 5.2.1.

Example 5.2.1. Let $Y = \{0, \alpha\} = K_{1,1}$ with $0 < \alpha$ and let S be a strong semilattice of left groups $T_0 = L_2 \times \mathbb{Z}_2$ and $T_\alpha = L_1 \times \mathbb{Z}_3$ with $\varphi_{\alpha,0} = c_{\alpha,(l_1,0)_0}$ such that $End(\mathbb{Z}_2)$ and $End(\mathbb{Z}_3)$ are idempotent-closed (see Example 1.3.2). Then the monoid $End(S)$ is idempotent-closed by Corollary 5.2.1, the figure is shown as follows.

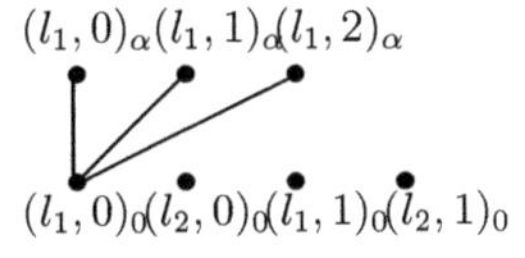

If we take $T_0 = L_3 \times \mathbb{Z}_2$ and $T_\alpha = L_1 \times \mathbb{Z}_3$, then the monoid $End(S)$ is not idempotent-closed since $n_0 = 3$. To see this, we take idempotents $f, g \in End(S)$ as follows.

$$f = \begin{pmatrix} (l_1,0)_0 & (l_2,0)_0 & (l_3,0)_0 & (l_1,1)_0 & (l_2,1)_0 & (l_3,1)_0 & (l_1,0)_\alpha & (l_1,1)_\alpha & (l_1,2)_\alpha \\ (l_1,0)_0 & (l_3,0)_0 & (l_3,0)_0 & (l_1,1)_0 & (l_3,1)_0 & (l_3,1)_0 & (l_3,0)_0 & (l_3,0)_0 & (l_3,0)_0 \end{pmatrix}$$

and

$$g = \begin{pmatrix} (l_1,0)_0 & (l_2,0)_0 & (l_3,0)_0 & (l_1,1)_0 & (l_2,1)_0 & (l_3,1)_0 & (l_1,0)_\alpha & (l_1,1)_\alpha & (l_1,2)_\alpha \\ (l_2,0)_0 & (l_2,0)_0 & (l_3,0)_0 & (l_2,1)_0 & (l_2,1)_0 & (l_3,1)_0 & (l_2,0)_0 & (l_2,0)_0 & (l_2,0)_0 \end{pmatrix}$$

but gf is not idempotent because $gfgf((l_1,0)_0) = (l_3,0)_0$ while $gf((l_1,0)_0) = (l_2,0)_0$.

5.3 Orthodox monoids

In this section we consider the strong semilattices of left groups whose endomorphism monoids are orthodox.

Corollary 5.3.1. *Let* $S = [Y; L_{n_\alpha} \times G_\alpha, c_{\alpha,(l_\beta,e_\beta)}]$ *be a non-trivial strong semilattice of left group* $L_{n_\alpha} \times G_\alpha$*. Then the monoid* $End(S)$ *is orthodox if and only if the following conditions hold*

1) $Y = Y_{0,m}$, $n_\xi = 2$,

2) $|Hom(G_0, G_\alpha)| = 1$, $n_0 = 1$ *and*

3) the monoid $End(G_\alpha)$ *is idempotent-closed, and*

4) the set $Hom(G_\alpha, G_\beta)$ *is hom-regular for all* $\alpha, \beta \in Y$.

Proof. See Corollaries 5.1.6 and 5.2.1. □

Now the defining homomorphisms are bijective.

Corollary 5.3.2. *Let* $S = [Y; L_{n_\alpha} \times G_\alpha, \varphi_{\alpha,\beta}]$ *be a non-trivial strong semilattice of left groups* $L_{n_\alpha} \times G_\alpha$ *with bijective* $\varphi_{\alpha,\beta}$*. Then the monoid* $End(S)$ *is orthodox if and only* $Y = Y_{0,m}$, $n_\alpha = 2$ *and the* $End(G)$ *is orthodox.*

Proof. See Corollaries 5.1.8 and 5.2.2. □

The following example shows positive and negative for Corollary 5.3.2.

Example 5.3.1. Let $Y = \{0, \alpha\} = K_{1,1}$ with $0 < \alpha$ and let S be a strong semilattice of left groups T_0 and T_α with bijective defining homomorphism $\varphi_{\alpha,0}$ where $T_0 = T_\alpha = L_2 \times \mathbb{Z}_6$ and the monoid $End(\mathbb{Z}_6)$ is regular and orthodox since $(End(\mathbb{Z}_6), \circ) \cong (\mathbb{Z}_6, \cdot)$ where $\cdot$ is the usual multiplication. (see Example 1.3.2). Then the monoid $End(S)$ is orthodox by Corollary 5.3.2, the figure is shown as follows.

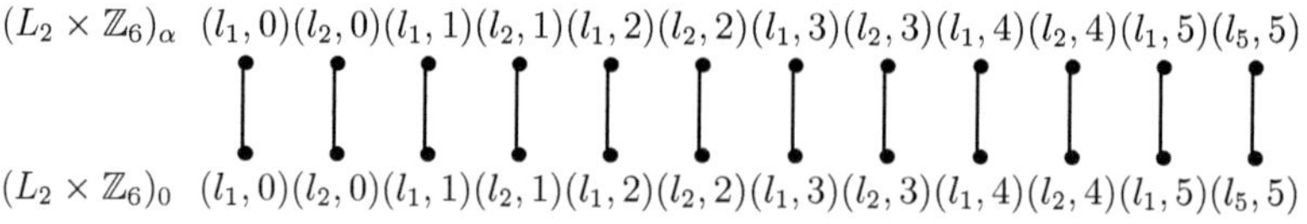

If we take $L_2 \times \mathbb{Z}_6$ by $L_2 \times \mathbb{Z}_4$, then the monoid $End(S)$ is not orthodox (see also Example 1.3.2) since the monoid $End(\mathbb{Z}_4)$ is not regular, so $End(L_2 \times \mathbb{Z}_4)$ is also not regular.

5.4 Left inverse endomorphisms

In this section we consider the strong semilattices of left groups whose endomorphism monoids are left inverse.

Corollary 5.4.1. *Let* $S = [Y; L_{n_\alpha} \times G_\alpha, c_{\alpha,(l_\beta,e_\beta)}]$ *be a non-trivial strong semilattice of left groups* $L_{n_\alpha} \times G_\alpha$. *Then the monoid End(S) is left inverse if and only if* $Y = Y_{0,m}$, $n_\xi = 2$ *and the monoid* $End(G_\xi)$ *is left inverse for each* $\xi \in Y_{0,n}$.

Proof. See Theorem 3.5.1 and Lemma 5.1.2. □

Now the defining homomorphisms are bijective.

Corollary 5.4.2. *Let* $S = [Y; L_{n_\alpha} \times G_\alpha, \varphi_{\alpha,\beta}]$ *be a non-trivial strong semilattice of left groups* $L_{n_\alpha} \times G_\alpha$ *with isomorphisms* $\varphi_{\alpha,\beta}$. *Then the monoid End(S) is left inverse if and only if* $Y = Y_{0,m}$, $n_\alpha = 2$ *and End(G) is left inverse.*

Proof. See Theorem 3.5.2 and Lemma 5.1.2. □

5.5 Completely regular monoids

In this section we consider the strong semilattices of left groups whose endomorphism monoids are completely regular.

Corollary 5.5.1. *Let* $S = [Y; L_{n_\alpha} \times G_\alpha, c_{\alpha,(l_\beta,e_\beta)}]$ *be a non-trivial strong semilattice of left groups* $L_{n_\alpha} \times G_\alpha$ *with* $\nu = \wedge Y$. *Then the monoid End(S) is completely regular if and only if if the following conditions hold*

1) $|Y| = 2$, $n_\xi = 2$,

2) $|Hom(G_\nu, G_\alpha)| = 1$, $n_\nu = 1$ *and*

3) the monoid $End(G_\xi)$ *is completely regular for each* $\xi \in Y$.

Proof. See Theorems 3.6.1, 3.6.2 and Lemma 5.1.2. □

Now the defining homomorphisms are bijective.

Corollary 5.5.2. *Let* $S = [Y; L_{n_\alpha} \times G_\alpha, \varphi_{\alpha,\beta}]$ *be a non-trivial strong semilattice of left groups* $L_{n_\alpha} \times G_\alpha$ *with bijective* $\varphi_{\alpha,\beta}$ *and* $\nu = \wedge Y$. *Then the monoid End(S) is completely regular if and only if* $|Y| = 2$, $n_\alpha = 2$ *and the monoid End(G) is completely regular.*

Proof. See Theorem 3.6.3 and Lemma 5.1.2. □

The following example shows positive and negative examples for Corollary 5.5.2.

The following example shows positive and negative examples for Corollary 5.5.2.

Example 5.5.1. We take the Example 5.3.1. Let $Y = \{0, \alpha\} = K_{1,1}$ with $0 < \alpha$ and let S be a strong semilattice of left groups T_0 and T_α such that $T_0 = T_\alpha = L_2 \times \mathbb{Z}_6$ with bijective defining homomorphism $\varphi_{\alpha,0}$. The set of endomorphisms of $End(\mathbb{Z}_6)$ is shown in Example 5.3.1. Take any endomorphisms $f(1) = 2$ and $g(1) = 5$ such that $fff = f$ and $ggg = g$. The monoid $End(\mathbb{Z}_6)$ is completely regular, and therefore the monoid $End(S)$ is completely regular by Corollary 5.5.2.

If we replace $\mathbb{Z}_6$ by $\mathbb{Z}_2 \times \mathbb{Z}_2$ such that $End(\mathbb{Z}_2 \times \mathbb{Z}_2)$ is not completely regular by calculating, and therefore the monoid $End(S)$ is not completely regular.

5.6 Idempotent monoids

In this section we consider the strong semilattices of left groups whose endomorphism monoids are idempotent.

Corollary 5.6.1. *Let $S = [Y; L_{n_\alpha} \times G_\alpha, c_{\alpha,(l_\beta,e_\beta)}]$ be a non-trivial strong semilattice of left groups $L_{n_\alpha} \times G_\alpha$ with $\nu = \wedge Y$. Then the monoid $End(S)$ is idempotent if and only if the following conditions hold*

1) $Y = \{\nu, \mu\}$ with $\nu < \mu$, $n_\xi = 1$ for all $\xi \in Y$,

2) $G_\nu, G_\mu \in \{\mathbb{Z}_1, \mathbb{Z}_2\}$, $G_\nu \neq G_\mu$, and

Proof. See Theorem 3.7.1 and Lemma 5.1.2. □

Now the defining homomorphisms are bijective.

Corollary 5.6.2. *Let $S = [Y; L_{n_\alpha} \times G_\alpha, \varphi_{\alpha,\beta}]$ be a non-trivial strong semilattice of left groups $L_{n_\alpha} \times G_\alpha$ with bijective $\varphi_{\alpha,\beta}$ and $\nu = \wedge Y$. Then the monoid $End(S)$ is idempotent if and only if $|Y| = 2$, $n_\alpha = 1$ and $G \in \{\mathbb{Z}_1, \mathbb{Z}_2\}$.*

Proof. See Theorem 3.7.2 and Lemma 5.1.2. □

Chapter 6

Generalization to surjective defining homomorphisms

In Chapter 2 we presented strong semilattices of left simple semigroups with constant defining homomorphisms and isomorphisms whose endomorphism monoids are regular, idempotent-closed, orthodox, left inverse, completely regular and idempotent.

In this chapter we consider strong semilattices of left simple semigroups with surjective defining homomorphisms whose endomorphism monoids have such properties. In this chapter we mainly consider the semilattice $Y = Y_{0,n}$.

6.1 Regular monoids

In this section we study strong semilattices of left simple semigroups whose endomorphism monoids are regular.

Construction 6.1.1. *Let $Y = Y_{0,n}$ and $S = [Y_{0,n}; S_\alpha, e_\alpha, \varphi_{\alpha,\beta}]$ be a non-trivial strong semilattice of left simple semigroup with surjective defining homomorphisms $\varphi_{\alpha,\beta}$. Take $\alpha, \beta \in Y_{0,n}$, $\alpha, \beta \neq 0$ and take $f_\alpha \in Hom(S_\alpha, S_\beta)$. Define $f : S \to S$ as follows*

$$f(x_\xi) := \begin{cases} f_\alpha(x_\alpha) \in S_\beta & \text{if } \xi = \alpha, \\ \varphi_{\beta,0}(f_\alpha(y_\alpha)) \in S_0 & \text{if } \xi \neq \alpha,\ \varphi_{\alpha,0}(y_\alpha) = \varphi_{\xi,0}(x_\xi), \end{cases}$$

for every $x_\xi \in S, \xi \in Y_{0,n}$. Then $f \in End(S)$.

Proof. It can be checked that f is well-defined. We show that f is a homomorphism.

Take $x_\gamma, y_\delta \in S,\ \gamma, \delta \in Y_{0,n}$.

Case 1. $\gamma, \delta = \alpha$. Thus

$$f(x_\alpha y_\alpha) = f_\alpha(x_\alpha y_\alpha) = f_\alpha(x_\alpha) f_\alpha(y_\alpha) = f(x_\alpha) f(y_\alpha).$$

Case 2. $\gamma = \alpha, \delta \neq \alpha$. Thus

$$\begin{aligned} f(x_\alpha y_\delta) &= f(\varphi_{\alpha,0}(x_\alpha)\varphi_{\delta,0}(y_\delta)) \\ &= \varphi_{\beta,0}(f_\alpha(z_\alpha)) \text{ where } \varphi_{\alpha,0}(z_\alpha) = \varphi_{\alpha,0}(x_\alpha)\varphi_{\delta,0}(y_\delta) \end{aligned}$$

and

$$\begin{aligned} f(x_\alpha) f(y_\delta) &= f_\alpha(x_\alpha)\varphi_{\beta,0}(f_\alpha(w_\alpha)) \text{ where } \varphi_{\alpha,0}(w_\alpha) = \varphi_{\delta,0}(y_\delta) \\ &= \varphi_{\beta,0}(f_\alpha(x_\alpha))\varphi_{\beta,0}(f_\alpha(w_\alpha)) \\ &= \varphi_{\beta,0}(f_\alpha(x_\alpha w_\alpha)) \end{aligned}$$

where $\varphi_{\alpha,0}(z_\alpha) = \varphi_{\alpha,0}(x_\alpha)\varphi_{\delta,0}(y_\delta) = \varphi_{\alpha,0}(x_\alpha)\varphi_{\alpha,0}(w_\alpha) = \varphi_{\alpha,0}(x_\alpha w_\alpha)$.

Case 3. $\gamma \neq \alpha, \delta \neq \alpha$. Thus

$$\begin{aligned} f(x_\gamma y_\delta) &= f(\varphi_{\gamma,0}(x_\gamma)\varphi_{\delta,0}(y_\delta)) \\ &= \varphi_{\beta,0}(f_\alpha(z_\alpha)) \text{ where } \varphi_{\alpha,0}(z_\alpha) = \varphi_{\gamma,0}(x_\gamma)\varphi_{\delta,0}(y_\delta) \end{aligned}$$

and

$$\begin{aligned} f(x_\gamma) f(y_\delta) &= \varphi_{\beta,0}(f_\alpha(t_\alpha))\varphi_{\beta,0}(f_\alpha(w_\alpha)) \text{ where } \varphi_{\alpha,0}(t_\alpha) = \varphi_{\gamma,0}(x_\gamma),\ \varphi_{\alpha,0}(w_\alpha) = \varphi_{\delta,0}(y_\delta) \\ &= \varphi_{\beta,0}(f_\alpha(t_\alpha w_\alpha)), \end{aligned}$$

where $\varphi_{\alpha,0}(z_\alpha) = \varphi_{\gamma,0}(x_\gamma)\varphi_{\delta,0}(y_\delta) = \varphi_{\alpha,0}(t_\alpha)\varphi_{\alpha,0}(w_\alpha) = \varphi_{\alpha,0}(t_\alpha w_\alpha)$.

Thus $f \in End(S)$. □

Lemma 6.1.1. *Let* $S = [Y_{0,n}; S_\alpha, e_\alpha, \varphi_{\alpha,\beta}]$ *be a non-trivial strong semilattice of left simple semigroup with surjective defining homomorphisms* $\varphi_{\alpha,\beta}$. *If the monoid* $End(S)$ *is regular, then the set* $Hom(S_\alpha, S_\beta))$ *is hom-regular for all* $\alpha, \beta \in Y$.

Proof. We remark first that since the defining homomorphisms $\varphi_{\alpha,\beta}$ are surjective, for each $y_\alpha \in S_\alpha$ there exists $x_\beta \in S,\ \beta < \alpha \in Y$ such that $\varphi_{\alpha,\alpha\beta}(y_\alpha) = \varphi_{\beta,\alpha\beta}(x_\beta)$.

Let $\alpha, \beta \in Y$. We show that the set $Hom(S_\alpha, S_\beta))$ is regular. Take $f_\alpha \in Hom(S_\alpha, S_\beta)$. We will define f which depends on α, β.

Case 1. $\alpha = 0, \beta \neq 0$, i.e., $f_0 \in Hom(S_0, S_\beta)$. Take $s \in End(Y_{0,n})$ such that $s(\xi) = \beta$ for all $\xi \in Y_{0,n}$. Using Lemma 3.1.1, for each $x_\xi \in S, \xi \in Y$, take $f \in End(S)$ as follows

$$f(x_\xi) := \begin{cases} f_0(x_0) \in S_\beta & \text{if } \xi = 0, \\ f_0(\varphi_{\xi,0}(x_\xi)) \in S_\beta & \text{if } \xi \neq 0, \end{cases}$$

By hypothesis there exists $f' \in End(S)$ such that $ff'f = f$. Thus

$$f_0(x_0) = f(x_0) = ff'f(x_0) = f_\gamma f'_\beta f_0(x_0)$$

where $f'_\beta \in Hom(S_\beta, S_\gamma)$ such that $\gamma \in \underline{f}^{-1}\{\beta\}$ and $f_\gamma \in Hom(S_\gamma, S_\beta)$. But by the definition of f, $Im(f_0) = Im(f_\alpha)$ for all $0 \neq \alpha \in Y$, so that γ may be 0.

Case 2. $\alpha \neq 0, \beta = 0$, i.e., $f_\alpha \in Hom(S_\alpha, S_0)$. In this case we can construct $f \in End(S)$ and $f(x_\alpha) = f_0(\varphi_{\alpha,0}(x_\alpha))$, i.e., f_α is determined by each $f_0 \in End(S_0)$ and we have $End(S_0)$ is regular.

Case 3. $\alpha, \beta \neq 0$, i.e., $f_\alpha \in Hom(S_\alpha, S_\beta)$. Take $s \in End(Y_{0,n})$ such that $s(\alpha) = \beta$ and $s(\gamma) = 0$ for all $\alpha \neq \gamma \in Y_{0,n}$. Using Construction 6.1.1, for each $x_\xi \in S, \xi \in Y$, take $f \in End(S)$ as follows

$$f(x_\xi) := \begin{cases} f_\alpha(x_\alpha) \in S_\beta & \text{if } \xi = \alpha, \\ \varphi_{\beta,0}(f_\alpha(z_\alpha)) \in S_0 & \text{if } \xi \neq \alpha,\ \varphi_{\alpha,0}(z_\alpha) = \varphi_{\xi,0}(x_\xi). \end{cases}$$

By hypothesis there exists $f' \in End(S)$ such that $ff'f = f$. Thus

$$f_\alpha(x_\alpha) = f(x_\alpha) = ff'f(x_\alpha) = f_\alpha f'_\alpha f_\alpha(x_\alpha)$$

where $f'_\alpha \in Hom(S_\beta, S_\alpha)$ because $\underline{f}^{-1}\{\beta\} = \{\alpha\}$ and $f_\alpha \in Hom(S_\alpha, S_\beta)$.

Therefore the set $Hom(S_\alpha, S_\beta)$ is hom-regular for $\alpha, \beta \in Y_{0,n}$. □

The converse is also true.

Lemma 6.1.2. *Let $Y = Y_{0,n}$ and let $S = [Y_{0,n}; S_\alpha, e_\alpha, \varphi_{\alpha,\beta}]$ be a non-trivial strong semilattice of left simple semigroup with surjective defining homomorphisms $\varphi_{\alpha,\beta}$. If the set $Hom(S_\alpha, S_\beta))$ is hom-regular for all $\alpha, \beta \in Y$, then the monoid $End(S)$ is regular.*

Proof. Take $f \in End(S)$. By Corollary 3.1.1 $\underline{f} \in End(Y_{0,n})$.

Case 1. $\underline{f}(\xi) = 0$ for all $\xi \in Y_{0,n}$. In this case $f_0 \in End(S_0)$, there exists $f'_0 \in End(S_0)$ such that $f_0 f'_0 f_0 = f_0$ and $f_\alpha(x_\alpha) = \varphi_{\underline{f}(\alpha),\underline{f}(0)} f_\alpha(x_\alpha) = f_0(\varphi_{\alpha,0}(x_\alpha))$ for all $\alpha \in Y_{n,0}$, $\alpha \neq 0$. Using Construction 3.3.1, for every $x_\xi \in S$, $\xi \in Y_{0,n}$, take $f' \in End(S)$ as follows

$$f'(x_\xi) := \begin{cases} f'_0(x_0) & \text{if } \xi = 0, \\ f'_0(\varphi_{\xi,0}(x_\xi)) & \text{if } \xi \neq 0. \end{cases}$$

Thus $ff'f = f$.

Case 2. $\underline{f}(\xi) = \alpha$ for all $\xi \in Y_{0,n}$ and for some $0 \neq \alpha \in Y_{0,n}$. Thus $f_\alpha \in End(S_\alpha)$, there exists $f'_\alpha \in End(S_\alpha)$ such that $f_\alpha f'_\alpha f_\alpha = f_\alpha$ since $End(S_\alpha)$ is regular by hypothesis. Using Construction 6.1.1, for each $x_\xi \in S, \xi \in Y_{0,n}$, take $f' \in End(S)$ as follows

$$f'(x_\xi) := \begin{cases} f'_\alpha(x_\alpha) \in S_\beta & \text{if } \xi = \alpha, \\ \varphi_{\beta,0}(f'_\alpha(z_\alpha)) \in S_0 & \text{if } \xi \neq \alpha,\ \varphi_{\alpha,0}(z_\alpha) = \varphi_{\xi,0}(x_\xi). \end{cases}$$

Thus $ff'f = f$.

Case 3. $\underline{f}$ is not constant. Thus $\underline{f}(0) = 0$ and for some $\alpha \neq 0$ we have $\underline{f}(\alpha) = \alpha$ or $\underline{f}(\alpha) = \beta$ for some $\beta \neq \alpha$. Further, in this case we have $f_\beta(x_\beta) = f_0(\varphi_{\beta,0}(x_\beta))$. In this case $f_0 \in End(S_0)$ which is regular, there exists $f'_0 \in End(S_0)$ such that $f_0 f'_0 f_0 = f_0$. If $f_\alpha \in End(S_\alpha)$ there exists $f'_\alpha \in End(S_\alpha)$ such that $f_\alpha f'_\alpha f_\alpha = f_\alpha$. If $f_\alpha \in Hom(S_\alpha, S_\beta)$ there exists $f'_\alpha \in Hom(S_\beta, S_\alpha)$ such that $f_\alpha f'_\alpha f_\alpha = f_\alpha$. Using Construction 6.1.1, for each $x_\xi \in S, \xi \in Y_{0,n}$, take $f' \in End(S)$ as follows

$$f'(x_\xi) := \begin{cases} f'_\alpha(x_\alpha) & \text{if } \xi = \alpha, \\ f'_0(x_0) & \text{if } \xi = 0, \\ \varphi_{\beta,0}(f'_\alpha(z_\alpha)) \in S_0 & \text{if } \xi \neq \alpha,\ \varphi_{\alpha,0}(z_\alpha) = \varphi_{\xi,0}(x_\xi). \end{cases}$$

Thus $ff'f = f$. □

Theorem 6.1.1. *Let $S = [Y_{0,n}; S_\alpha, e_\alpha, \varphi_{\alpha,\beta}]$ be a non-trivial strong semilattice of left simple semigroup with surjective defining homomorphisms $\varphi_{\alpha,\beta}$. Then the monoid $End(S)$ is regular if and only if the set $Hom(S_\alpha, S_\beta)$ is hom-regular for all $\alpha, \beta \in Y$.*

Problem 6.1.1. Find the conditions when the semilattice more general than $Y_{0,m}$.

6.2 Idempotent-closed monoids

In this section we consider strong semilattices of left simple semigroups whose endomorphism monoids are idempotent-closed.

Lemma 6.2.1. *Let $S = [Y_{0,n}; S_\alpha, e_\alpha, \varphi_{\alpha,\beta}]$ be a non-trivial strong semilattice of left simple semigroups with surjective defining homomorphisms $\varphi_{\xi,0}$. Let $f \in End(S)$ and let $\alpha \in Y_{0,n}$. Then the following hold*

1) If $\underline{f}(\xi) = \alpha$, then $f_0(\varphi_{\alpha,0}(x_\alpha)) = f_\alpha(x_\alpha) = f_\beta(y_\beta) = f_0(\varphi_{\beta,0}(y_\beta))$ for all $y_\beta \in G_\beta$ such that $\varphi_{\beta,0}(y_\beta) = \varphi_{\alpha,0}(x_\alpha)$.

In particular, if f is idempotent, then $f_\alpha(x_\alpha) = f_0(\varphi_{\alpha,0}(x_\alpha)) = (f_0(\varphi_{\alpha,0}))^2(x_\alpha) = (f_\alpha)^2(x_\alpha)$.

2) Let $x_\alpha, y_\beta \in S,\ \alpha, \beta \in Y_{0,n}$ be such that $\varphi_{\alpha,0}(x_\alpha) = \varphi_{\beta,0}(y_\beta)$. Then $\varphi_{\underline{f}(\alpha),\underline{f}(0)}(f_\alpha(x_\alpha)) = \varphi_{\underline{f}(\beta),\underline{f}(0)}(f_\beta(y_\beta))$.

Proof. 1) We have $f_0\varphi_{\alpha,0}(x_\alpha) = \varphi_{\underline{f}(\alpha),\underline{f}(0)}f_\alpha(x_\alpha) = \varphi_{\alpha,\alpha}(f_\alpha(x_\alpha)) = f_\alpha(x_\alpha)$ where $f_0 \in Hom(S_0, S_\alpha)$ and $f_\alpha \in End(S_\alpha)$. Since $\varphi_{\alpha,0}$ is surjective, there exists $y_\beta \in S_\beta,\ \beta \in Y_{0,n}$ with $0 \neq \alpha \neq \beta$ such that $\varphi_{\alpha,0}(x_\alpha) = \varphi_{\beta,0}(y_\beta)$. Then

$$\begin{aligned} f_0(\varphi_{\alpha,0}(x_\alpha)) &= f_0(\varphi_{\beta,0}(y_\beta)) \\ &= \varphi_{\underline{f}(\beta),\underline{f}(0)}f_\beta(y_\beta) \\ &= \varphi_{\alpha,\alpha}f_\beta(y_\beta) \\ &= f_\beta(y_\beta). \end{aligned}$$

Thus we have $f_0(\varphi_{\alpha,0}(x_\alpha)) = f_\alpha(x_\alpha) = f_\beta(y_\beta) = f_0(\varphi_{\beta,0}(y_\beta))$ for all $y_\beta \in S_\beta$ such that $\varphi_{\beta,0}(y_\beta) = \varphi_{\alpha,0}(x_\alpha)$.

If f is idempotent, we have $ff(x_\xi) = f(x_\xi)$ for all $x_\xi \in S,\ \xi \in Y_{0,n}$. Thus

$$\begin{aligned} f_0(\varphi_{\alpha,0}(x_\alpha)) &= f_\alpha(x_\alpha) \\ &= f(x_\alpha) \\ &= ff(x_\alpha) \\ &= f_\alpha f_\alpha(x_\alpha) \\ &= f_0\varphi_{\alpha,0}(f_0(\varphi_{\alpha,0}(x_\alpha)) \\ &= (f_0\varphi_{\alpha,0})^2(x_\alpha). \end{aligned}$$

Then $f_\alpha = f_0\varphi_{\alpha,0} \in End(S_\alpha)$ for every $x_\alpha \in S_\alpha$.

2) Since $f \in End(S)$, we have $f(x_\alpha e_0) = f(x_\alpha)f(e_0)$. Then

$$\begin{aligned} f(x_\alpha e_0) &= f(\varphi_{\alpha,0}(x_\alpha)e_0) \\ &= f_0(\varphi_{\alpha,0}(x_\alpha)) \\ &= f_0(\varphi_{\beta,0}(y_\beta))\ (\text{since } \varphi_{\alpha,0}(x_\alpha) = \varphi_{\beta,0}(y_\beta)) \\ &= \varphi_{\underline{f}(\beta),\underline{f}(0)}(f_\beta(x_\beta)) \end{aligned}$$

and

$$\begin{aligned} f(x_\alpha)f(e_0) &= f_\alpha(x_\alpha)f_0(e_0) \\ &= \varphi_{\underline{f}(\alpha),\underline{f}(0)}(f_\alpha(x_\alpha))\varphi_{\underline{f}(0),\underline{f}(0)}(f_0(e_0)) \\ &= \varphi_{\underline{f}(\alpha),\underline{f}(0)}(f_\alpha(x_\alpha)). \end{aligned}$$

This implies that $\varphi_{\underline{f}(\alpha),\underline{f}(0)}(f_\alpha(x_\alpha)) = \varphi_{\underline{f}(\beta),\underline{f}(0)}(f_\beta(y_\beta))$. □

Lemma 6.2.2. *Let $S = [Y_{0,n}; S_\alpha, e_\alpha, \varphi_{\alpha,\beta}]$ be a non-trivial strong semilattice of left simple semigroups with surjective defining homomorphisms $\varphi_{\alpha,\beta}$. If the monoid $End(S)$ is idempotent-closed, then the monoid $End(S_\xi)$ is idempotent-closed for every $\xi \in Y_{0,n}$.*

Proof. By Lemma 3.2.3 *End*(Y) is idempotent-closed. This implies that $Y = Y_{0,n}$ by Proposition 2.2.1.

We next show that *End*(S_ξ) is idempotent-closed for $\xi \in Y_{0,n}$.

Case 1. We show that *End*(S_0) is idempotent-closed, take two idempotents $f_0, h_0 \in$ *End*(S_0). Using Construction 3.3.1, take $f, h \in$ *End*(S) as follows

$$f(x_\alpha) := \begin{cases} f_0(x_0) & \text{if } \alpha = 0, \\ f_0(\varphi_{\alpha,0}(x_\alpha)) & \text{if } \alpha \neq 0, \end{cases}$$

and

$$h(x_\alpha) := \begin{cases} h_0(x_0) & \text{if } \alpha = 0, \\ h_0(\varphi_{\alpha,0}(x_\alpha)) & \text{if } \alpha \neq 0, \end{cases}$$

for every $x_\alpha \in S,\ \alpha \in Y_{0,n}$. Then f, h are idempotents. By hypothesis fh is idempotent. For each $x_0 \in S_0$, we have $f_0h_0f_0h_0(x_0) = fhfh(x_0) = fh(x_0) = f_0h_0(x_0)$. This implies f_0h_0 is idempotent and therefore *End*(S_0) is idempotent-closed.

Case 2. We show that *End*(S_α) is idempotent-closed, $0 \neq \alpha \in Y_{0,n}$, take two idempotents $f_\alpha, h_\alpha \in$ *End*(S_α). We note that for each $x_0 \in S_0$, there exists $y_\alpha \in S_\alpha,\ 0 \neq \alpha \in Y_{n,0}$ such that $\varphi_{\alpha,0}(y_\alpha) = x_0$. Using Construction 6.1.1, for every $x_\xi \in S,\ \xi \in Y_{0,n}$, take $f, h \in$ *End*(S) as follows

$$f(x_\xi) := \begin{cases} f_\alpha(x_\alpha) & \text{if } \xi = \alpha, \\ \varphi_{\beta,0}(f_\alpha(z_\alpha)) & \text{if } \xi \neq \alpha \text{ and } \varphi_{\alpha,0}(z_\alpha) = \varphi_{\xi,0}(x_\xi), \end{cases}$$

$$h(x_\xi) := \begin{cases} h_\alpha(x_\alpha) & \text{if } \xi = \alpha, \\ \varphi_{\beta,0}(h_\alpha(z_\alpha)) & \text{if } \xi \neq \alpha \text{ and } \varphi_{\alpha,0}(z_\alpha) = \varphi_{\xi,0}(x_\xi). \end{cases}$$

Then f, h are idempotents. By hypothesis fh is idempotent. Thus

$$\begin{aligned} f_\alpha h_\alpha f_\alpha h_\alpha(x_\alpha) &= fhfh(x_\alpha) \\ &= fh(x_\alpha) \\ &= f_\alpha(h_\alpha(x_\alpha)). \end{aligned}$$

Thus $f_\alpha h_\alpha$ is idempotent. Hence *End*(S_ξ) is idempotent-closed for all $\xi \in Y_{0,n}$. □

Lemma 6.2.3. *Let $S = [Y; S_\alpha, e_\alpha, \varphi_{\alpha,\beta}]$ be a non-trivial strong semilattice of left simple semigroups with surjective defining homomorphisms $\varphi_{\alpha,\beta}$. If the monoid End(S_ξ) is idempotent-closed for all $\xi \in Y_{0,n}$, then the monoid End(S) is idempotent-closed.*

Proof. Take two idempotents $f, h \in$ *End*(S). Then $\underline{f}, \underline{h} \in$ *End*$(Y_{0,n})$.

Case 1. If $\underline{f}$ (or $\underline{h}$) is constant. Suppose that $\underline{f}(\xi) = \alpha$ and $\underline{h}(\xi) = \beta$ for all $\xi \in Y_{0,n}$. Then $\underline{f}\ \underline{h}(\xi) = \alpha$. For every $x_\xi \in S,\ \xi \in Y_{0,n}$, we have

$$\begin{aligned}
fhfh(x_\xi) &= fhfh_\xi(x_\xi) \\
&= fhf(h_0(\varphi_{\xi,0}(x_\xi))) \\
&= fhf(h_0(\varphi_{\alpha,0}(y_\alpha))) \text{ where } (\varphi_{\xi,0}(x_\xi) = \varphi_{\alpha,0}(y_\alpha)) \\
&= f_\beta h_\alpha f_\beta(h_0(\varphi_{\alpha,0}(y_\alpha))) \\
&= f_0\varphi_{\beta,0}h_0\varphi_{\alpha,0}f_0\varphi_{\beta,0}(h_0(\varphi_{\alpha,0}(y_\alpha))) \\
&= (f_0\varphi_{\beta,0}h_0\varphi_{\alpha,0})^2(y_\alpha))) \\
&= f_0\varphi_{\beta,0}h_0\varphi_{\alpha,0}(y_\alpha))) \\
&= fh(x_\xi)
\end{aligned}$$

where $f_0\varphi_{\beta,0}h_0\varphi_{\alpha,0} \in End(S_\alpha)$. Thus fh is idempotent.

Case 2. If $\underline{f}$ and $\underline{h}$ are not constants. Thus $\underline{f}\ \underline{h}(\alpha) = \alpha$ if $\underline{f}(\alpha) = \alpha$ and $\underline{h}(\alpha) = \alpha$. Further, in this case we have $f_\alpha(x_\alpha) = f(x_\alpha) = ff(x_\alpha) = f_\alpha f_\alpha(x_\alpha)$ and $h_\alpha(x_\alpha) = h(x_\alpha) = hh(x_\alpha) = h_\alpha h_\alpha(x_\alpha)$, i.e., f_α, h_α are idempotents. This implies that $f_\alpha h_\alpha \in End(S_\alpha)$ which is also idempotent, since $End(S_\alpha)$ is idempotent-closed. Thus

$$fhfh(x_\alpha) = f_\alpha h_\alpha f_\alpha h_\alpha(x_\alpha) = f_\alpha h_\alpha(x_\alpha) = fh(x_\alpha).$$

In other case, we have $\underline{f}\ \underline{h}(\alpha) = 0$ where $\underline{f}(0) = \underline{h}(0) = 0$. Thus $f_0(x_0) = f(x_0) = ff(x_0) = f_0f_0(x_0)$ and $h_0(x_0) = h(x_0) = hh(x_0) = h_0h_0(x_0)$. Then f_0, h_0 are idempotents. This implies that $f_0h_0 \in End(S_0)$ which is also idempotent, since $End(S_0)$ is idempotent-closed. Then

$$fhfh(x_\alpha) = fhfh_\alpha(x_\alpha) = f_0h_0f_0h_0(\varphi_{\alpha,0}(x_\alpha)) = f_0h_0(\varphi_{\alpha,0}(x_\alpha)) = fh(x_\alpha).$$

□

In the next theorem we get from Lemmas 6.2.2 and 6.2.3.

Theorem 6.2.1. *Let $S = [Y_{0,n}; S_\alpha, e_\alpha, \varphi_{\alpha,\beta}]$ be a non-trivial strong semilattice of left simple semigroups with surjective defining homomorphisms $\varphi_{\alpha,\beta}$. Then the monoid $End(S)$ is idempotent-closed if and only if the monoid $End(S_\xi)$ is idempotent-closed for every $\xi \in Y_{0,n}$.*

Since groups are left simple semigroups, we have the next example to illustrate Theorem 6.2.1.

Example 6.2.1. Consider a three-element semilattice $Y_{0,2} = \{0 < \alpha, \beta\}$. Let $G_0 = \mathbb{Z}_2, G_\alpha = S_3$ and $G_\beta = \mathbb{Z}_4$ where $\mathbb{Z}_4, \mathbb{Z}_2$ are the additive groups modulo 4 and 2, respectively and S_3 is the symmetric group order 6. Their monoids $End(\mathbb{Z}_2)$, $End(\mathbb{Z}_4)$ and $End(S_3)$ are idempotent-closed (see Example 1.3.2). Take the strong semilattice of groups $S = G_0 \bigcup G_\alpha \bigcup G_\beta$ with the defining homomorphisms as shown below.

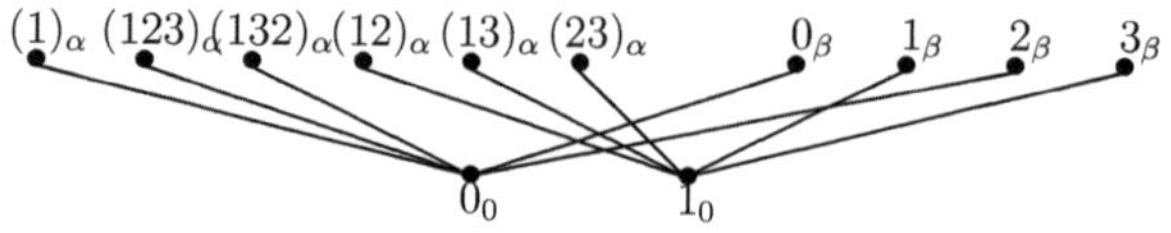

Take the two idempotents $f_\alpha = \begin{pmatrix} (1)_\alpha & (123)_\alpha & (132)_\alpha & (12)_\alpha & (13)_\alpha & (23)_\alpha \\ (1)_\alpha & (1)_\alpha & (1)_\alpha & (12)_\alpha & (12)_\alpha & (12)_\alpha \end{pmatrix}$ and

$$h_\alpha = \begin{pmatrix} (1)_\alpha & (123)_\alpha & (132)_\alpha & (12)_\alpha & (13)_\alpha & (23)_\alpha \\ (1)_\alpha & (1)_\alpha & (1)_\alpha & (23)_\alpha & (23)_\alpha & (23)_\alpha \end{pmatrix} \in End(G_\alpha).$$

Take the idempotent $s = \begin{pmatrix} 0 & \alpha & \beta \\ 0 & \alpha & 0 \end{pmatrix} \in End(Y)$ and for $0, \alpha \in Im(s)$, and let $f_0 = h_0$ be the identity map on $G_0 = \mathbb{Z}_2$. Then we can construct $f, h \in End(S)$ as in Construction 6.1.1, so we have

$$f = \begin{pmatrix} 0_0 & 1_0 & (1)_\alpha & (123)_\alpha & (132)_\alpha & (12)_\alpha & (13)_\alpha & (23)_\alpha & 0_\beta & 1_\beta & 2_\beta & 3_\beta \\ 0_0 & 1_0 & (1)_\alpha & (123)_\alpha & (132)_\alpha & (12)_\alpha & (12)_\alpha & (12)_\alpha & 0_0 & 1_0 & 0_0 & 1_0 \end{pmatrix}$$

and

$$h = \begin{pmatrix} 0_0 & 1_0 & (1)_\alpha & (123)_\alpha & (132)_\alpha & (12)_\alpha & (13)_\alpha & (23)_\alpha & 0_\beta & 1_\beta & 2_\beta & 3_\beta \\ 0_0 & 1_0 & (1)_\alpha & (123)_\alpha & (132)_\alpha & (23)_\alpha & (23)_\alpha & (23)_\alpha & 0_0 & 1_0 & 0_0 & 1_0 \end{pmatrix}.$$

Thus f and h are idempotents. It is clear that

$$fh = \begin{pmatrix} 0_0 & 1_0 & (1)_\alpha & (123)_\alpha & (132)_\alpha & (12)_\alpha & (13)_\alpha & (23)_\alpha & 0_\beta & 1_\beta & 2_\beta & 3_\beta \\ 0_0 & 1_0 & (1)_\alpha & (123)_\alpha & (132)_\alpha & (12)_\alpha & (12)_\alpha & (12)_\alpha & 0_0 & 1_0 & 0_0 & 1_0 \end{pmatrix}$$

is idempotent such that $fh(x_\alpha) = f_\alpha h_\alpha(x_\alpha)$ for every $x_\alpha \in G_\alpha$.

6.3 Orthodox monoids

Now we consider strong semilattices of left simple semigroups whose endomorphism monoids are orthodox.

The following theorem follows from Theorems 6.1.1 and 6.2.1.

Theorem 6.3.1. *Let $S = [Y_{0,n}; S_\alpha, e_\alpha, \varphi_{\alpha,\beta}]$ be a non-trivial strong semilattice of left simple semigroups with surjective defining homomorphisms $\varphi_{\alpha,\beta}$. Then the monoid $End(S)$ is orthodox if and only if the following conditions hold*

1) the monoid $End(S_\xi)$ is idempotent-closed for every $\xi \in Y_{0,n}$, and

2) the set $Hom(S_\alpha, S_\beta)$ is hom-regular for every $\alpha, \beta \in Y_{0,n}$.

6.4 Left inverse monoids

In this section we consider strong semilattices of left simple semigroups with surjective defining homomorphisms whose endomorphism monoids are left inverse.

Lemma 6.4.1. *Let $S = [Y_{0,n}; S_\alpha, e_\alpha, \varphi_{\alpha,\beta}]$ be a non-trivial strong semilattice of left simple semigroups with surjective defining homomorphisms $\varphi_{\alpha,\beta}$. If the monoid $End(S)$ is left inverse, then the monoid $End(S_\xi)$ is left inverse for each $\xi \in Y_{0,n}$.*

Proof. By Lemma 3.2.3, the monoid $End(Y)$ is left inverse. This implies $Y = Y_{0,n}$ by Proposition 2.2.1.

We show that $End(S_\xi)$ is left inverse for $\xi \in Y_{0,n}$.

Case 1. We show that $End(S_0)$ is left inverse. Take two idempotents $f_0, h_0 \in End(S_0)$. Using Construction 3.3.1, take $f, h \in End(S)$ as follows

$$f(x_\xi) = \begin{cases} f_0(x_0) & \text{if } \xi = 0 \\ f_0(\varphi_{\xi,0}(x_\xi)) & \text{if } \xi \neq 0, \end{cases}$$

and

$$h(x_\xi) = \begin{cases} h_0(x_0) & \text{if } \xi = 0 \\ h_0(\varphi_{\xi,0}(x_\xi)) & \text{if } \xi \neq 0, \end{cases}$$

for every $x_\xi \in S$, $\xi \in Y_{0,n}$. By hypothesis $fhf = fh$. For each $x_0 \in S_0$, we have $f_0h_0f_0(x_0) = fhf(x_0) = fh(x_0) = f_0h_0(x_0)$. This implies $f_0h_0f_0 = f_0h_0$, and therefore $End(S_0)$ is left inverse.

Case 2. We show that $End(S_\alpha)$ is left inverse, $0 \neq \alpha \in Y_{0,n}$. Take two idempotents $f_\alpha, h_\alpha \in End(S_\alpha)$. For each $x_0 \in S_0$, there exists $y_\alpha \in S_\alpha$ such that $\varphi_{\alpha,0}(y_\alpha) = x_0$ since $\varphi_{\alpha,0}$ is surjective. Using Construction 6.1.1, for every $x_\xi \in S$, $\xi \in Y_{0,n}$, take $f, h \in End(S)$ as follows

$$f(x_\xi) = \begin{cases} f_\alpha(x_\alpha) & \text{if } \xi = \alpha, \\ \varphi_{\alpha,0}(f_\alpha(y_\alpha))) & \text{if } \xi \neq \alpha \text{ and } \varphi_{\alpha,0}(y_\alpha) = \varphi_{\xi,0}(x_\xi), \end{cases}$$

and

$$h(x_\xi) = \begin{cases} h_\alpha(x_\alpha) & \text{if } \xi = \alpha, \\ \varphi_{\alpha,0}(h_\alpha(y_\alpha))) & \text{if } \xi \neq \alpha \text{ and } \varphi_{\alpha,0}(y_\alpha) = \varphi_{\xi,0}(x_\xi). \end{cases}$$

Then f, h are idempotents. By hypothesis, fh is idempotent. Then

$$\begin{aligned} f_\alpha h_\alpha f_\alpha(x_\alpha) &= fhf(x_\alpha) \\ &= fh(x_\alpha) \\ &= f_\alpha(h_\alpha(x_\alpha)). \end{aligned}$$

Thus $f_\alpha h_\alpha$ is left inverse. Hence $End(S_\xi)$ is left inverse for all $\xi \in Y_{0,n}$. □

Lemma 6.4.2. *Let* $S = [Y_{0,n}; S_\alpha, e_\alpha, \varphi_{\alpha,\beta}]$ *be a non-trivial strong semilattice of left simple semigroups with surjective defining homomorphisms* $\varphi_{\alpha,\beta}$. *If the monoid* $End(S_\xi)$ *is left inverse for each* $\xi \in Y_{0,n}$, *then the monoid* $End(S)$ *is left inverse.*

Proof. Take two idempotents $f, h \in End(S)$. Then $\underline{f}, \underline{h} \in End(S)$ which are idempotents.

Case 1. If $\underline{f}$ and $\underline{h}$) are constant.

1.1 $\underline{f}(\xi) = 0$ and $\underline{h}(\xi) = 0$. Then $f_0, h_0 \in End(S_0)$ and f_α, h_α are determined by f_0 and h_0 respectively for $0 \neq \alpha \in Y_{0,n}$. That is

$$\begin{aligned} f_0(\varphi_{\alpha,0}(x_\alpha)) &= \varphi_{\underline{f}(\alpha),\underline{f}(0)}(f_\alpha(x_\alpha)) \\ &= \varphi_{0,0}(f_\alpha(x_\alpha)) \\ &= f_\alpha(x_\alpha) \end{aligned}$$

and

$$\begin{aligned} h_0(\varphi_{\alpha,0}(x_\alpha)) &= \varphi_{\underline{h}(\alpha),\underline{h}(0)}(h_\alpha(x_\alpha)) \\ &= \varphi_{0,0}(h_\alpha(x_\alpha)) \\ &= h_\alpha(x_\alpha). \end{aligned}$$

By using that $End(S_0)$ is left inverse, so that $f_0 h_0 f_0 = f_0 h_0$. Thus

$$\begin{aligned} fhf(x_\xi) &= f_0 h_0 f_0(\varphi_{\xi,0}(x_\xi)) \\ &= f_0 h_0(\varphi_{\xi,0}(x_\xi)) \\ &= fh(x_\xi) \end{aligned}$$

1.2 $\underline{f}(\xi) = 0$ and $\underline{h}(\xi) = \alpha$ for some $0 \neq \alpha \in Y_{0,n}$. Then $f_0, h_0 \in End(S_0)$ and f_α, h_α are determined by f_0 and h_0 respectively for $0 \neq \alpha \in Y_{0,n}$. That is

$$\begin{aligned} f_0(\varphi_{\alpha,0}(x_\alpha)) &= \varphi_{\underline{f}(\alpha),\underline{f}(0)}(f_\alpha(x_\alpha)) \\ &= \varphi_{0,0}(f_\alpha(x_\alpha)) \\ &= f_\alpha(x_\alpha) \end{aligned}$$

and

$$\begin{aligned} h_0(\varphi_{\alpha,0}(x_\alpha)) &= \varphi_{\underline{h}(\alpha),\underline{h}(0)}(h_\alpha(x_\alpha)) \\ &= \varphi_{0,0}(h_\alpha(x_\alpha)) \\ &= h_\alpha(x_\alpha). \end{aligned}$$

By using that $End(S_0)$ is left inverse, so that $f_0(\varphi_{\alpha,0}h_0)f_0 = f_0(\varphi_{\alpha,0}h_0)$ where $\varphi_{\alpha,0}h_0 \in End(S_0)$. Thus

$$\begin{aligned} fhf(x_\xi) &= fhf_0(\varphi_{\xi,0}(x_\xi)) \\ &= fh_0f_0(\varphi_{\xi,0}(x_\xi)) \\ &= f_0(\varphi_{\alpha,0}(h_0f_0(\varphi_{\xi,0}(x_\xi)))) \\ &= f_0(\varphi_{\alpha,0}h_0)f_0(\varphi_{\xi,0}(x_\xi)) \\ &= f_0(\varphi_{\alpha,0}h_0)(\varphi_{\xi,0}(x_\xi)) \\ &= f_0(\varphi_{\alpha,0}(h(x_\xi))) \\ &= fh(x_\xi) \end{aligned}$$

1.3 $\underline{f}(\xi) = \alpha$ and $\underline{h}(\xi) = \beta$ for some $0 \neq \alpha, \beta \in Y_{0,n}$. Then $\underline{fhf} = \underline{fh}$ and we have $\varphi_{\alpha,0}f_0, \varphi_{\beta,0}h_0 \in End(S_0)$. Thus $(\varphi_{\alpha,0}f_0)(\varphi_{\beta,0}h_0)(\varphi_{\alpha,0}f_0) = (\varphi_{\alpha,0}f_0)(\varphi_{\beta,0}h_0)$. Thus

$$\begin{aligned} fhf(x_\xi) &= fhf_0(\varphi_{\xi,0}(x_\xi)) \\ &= fh_0\varphi_{\alpha,0}f_0(\varphi_{\xi,0}(x_\xi)) \\ &= f_0\varphi_{\beta,0}h_0\varphi_{\alpha,0}f_0(\varphi_{\xi,0}(x_\xi)) \\ &= (\varphi_{\alpha,0}f_0)(\varphi_{\beta,0}h_0)(\varphi_{\alpha,0}f_0)(\varphi_{\xi,0}(x_\xi)) \\ &= (\varphi_{\alpha,0}f_0)(\varphi_{\beta,0}h_0)(\varphi_{\xi,0}(x_\xi)) \\ &= f_0(\varphi_{\alpha,0}h_0)h(x_\xi) \\ &= f_0(\varphi_{\alpha,0}h_0)(\varphi_{\xi,0}(x_\xi)) \\ &= f_0(\varphi_{\alpha,0}(h(x_\xi))) \\ &= fh(x_\xi) \end{aligned}$$

Suppose that $\underline{f}(\xi) = \alpha$ for all $\xi \in Y_{0,n}$. Then $\underline{f}\ \underline{h}(\xi) = \alpha$. In fact, for every $x_\xi \in S,\ \xi \in Y_{0,n}\ fhf(x_\xi) = f(x_\xi) = fh(x_\xi)$.

Case 2. If $\underline{f}$ and $\underline{h}$ are not constant. Then for each $\alpha \in Y_{0,n}$, $\underline{f}\ \underline{h}(\alpha) = \alpha$ if $\underline{f}(\alpha) = \alpha$ and $\underline{h}(\alpha) = \alpha$. Thus $f_\alpha(x_\alpha) = f(x_\alpha) = ff(x_\alpha) = f_\alpha f_\alpha(x_\alpha)$ and $h_\alpha(x_\alpha) = h(x_\alpha) =$

$hh(x_\alpha) = h_\alpha h_\alpha(x_\alpha)$, i.e., f_α, h_α are idempotents. This implies that $f_\alpha h_\alpha f_\alpha = f_\alpha h_\alpha$ which is also idempotent, since $End(S_\alpha)$ is left inverse. Thus

$$fhf(x_\alpha) = f_\alpha h_\alpha f_\alpha(x_\alpha) = f_\alpha h_\alpha(x_\alpha) = fh(x_\alpha).$$

In other cases, we have $\underline{f}\ \underline{h}(\alpha) = 0$ where $\underline{f}(0) = \underline{h}(0) = 0$. Thus $f_0(x_0) = f(x_0) = ff(x_0) = f_0 f_0(x_0)$ and $h_0(x_0) = h(x_0) = hh(x_0) = h_0 h_0(x_0)$. Then f_0, h_0 are idempotents. This implies that $f_0 h_0 f_0 = f_0 h_0$ which is also idempotent, since $End(S_0)$ is left inverse. Then

$$fhf(x_\alpha) = fhf_\alpha(x_\alpha) = f_0 h_0 f_0(\varphi_{\alpha,0}(x_\alpha)) = f_0 h_0(\varphi_{\alpha,0}(x_\alpha)) = fh(x_\alpha).$$

□

In the next theorem we get from Lemmas 6.4.1 and 6.4.2.

Theorem 6.4.1. *Let $S = [Y_{0,n}; S_\alpha, e_\alpha, \varphi_{\alpha,\beta}]$ be a non-trivial strong semilattice of left simple semigroups with surjective defining homomorphisms $\varphi_{\alpha,\beta}$. Then the monoid $End(S)$ is left inverse if and only if the monoid $End(S_\xi)$ is left inverse for every $\xi \in Y_{0,n}$.*

6.5 Completely regular monoids

We now consider strong semilattices of left simple semigroups whose endomorphism monoids are completely regular.

Lemma 6.5.1. *Let $Y = \{\mu, \nu\}$ with $\nu < \mu$ and let $S = [Y; S_\alpha, e_\alpha, \varphi_{\alpha,\beta}]$ be a non-trivial strong semilattice of left simple semigroups with surjective defining homomorphisms $\varphi_{\alpha,\beta}$. Let $f \in End(S)$. If $\underline{f}(\mu) = \underline{f}(\nu)$, then $Im(f_\nu) = Im(f_\mu)$.*

Proof. We have $Im(f_\mu) \subseteq Im(f_\nu)$. Let $x \in Im(f_\nu)$. There exists $y_\nu \in S_\nu$ such that $x = f_\nu(y_\nu)$. Since $\varphi_{\mu,\nu}$ is surjective, there exists $z_\mu \in S_\mu$ such that $\varphi_{\mu,\nu}(z_\mu) = y_\nu$. Thus $x = f_\nu(y_\nu) = f_\nu(\varphi_{\mu,\nu}(z_\mu)) = \varphi_{\underline{f}(\mu),\underline{f}(\nu)} = f_\mu(z_\mu)$. This implies $x \in Im(f_\mu)$, and therefore $Im(f_\mu) = Im(f_\nu)$. □

Lemma 6.5.2. *Let $S = [Y; S_\alpha, e_\alpha, \varphi_{\alpha,\beta}]$ be a non-trivial strong semilattice of left simple semigroups with surjective defining homomorphisms $\varphi_{\alpha,\beta}$. If the monoid $End(S)$ is completely regular, then $|Y| = 2$ and the monoid $End(S_\xi)$ is completely regular for each $\xi \in Y$.*

Proof. By Lemma 3.2.3, the monoid $End(Y)$ is completely regular. This implies $|Y| = 2$ by Proposition 2.2.1. Assume $Y = \{\nu, \mu\}, \nu < \mu$.

We show that $End(S_\xi)$ is completely regular for $\xi \in Y$.

Case 1. We show that $End(S_\nu)$ is completely regular. Take $f_\nu \in End(S_\nu)$.

Take $s \in End(Y)$ with $s(\mu) = s(\nu) = \nu$. Using Construction 3.3.1, for every $x_\xi \in S,\ \xi \in Y$, take $f \in End(S)$ as follows

$$f(x_\xi) := \begin{cases} f_\nu(x_\nu) & \text{if } \xi = \nu, \\ f_\nu(\varphi_{\mu,\nu}(x_\mu)) & \text{if } \xi = \mu. \end{cases}$$

By hypothesis, there exists $f' \in End(S)$ such that $ff'f = f$ and $ff' = f'f$. That is $f'(f_\nu(x_\nu)) = f'f(x_\nu) = ff'(x_\nu)$ for each $x_\nu \in S_\nu$. This implies $f'|_{S_\nu} \in End(S_\nu)$. Thus $f_\nu f'_\nu f_\nu(x_\nu) = ff'f(x_\nu) = f(x_\nu) = f_\nu(x_\nu)$ and $f'_\nu f_\nu(x_\nu) = f_\nu f'_\nu(x_\nu)$ for each $x_\nu \in S_\nu$. This implies f_ν is completely regular and therefore $End(S_\nu)$ is completely regular.

Case 2. We show that $End(S_\mu)$ is completely regular, take $f_\mu \in End(S_\mu)$ and take $s \in End(Y)$ such that $s(\mu) = \mu,\ s(\nu) = \nu$. Let $x_\nu \in S_\nu$, there exists $y_\mu \in S_\mu$ such that $\varphi_{\mu,\nu}(y_\mu) = x_\nu$, since $\varphi_{\mu,\nu}$ is surjective. Using Construction 6.1.1, for every $x_\xi \in S,\ \xi \in Y$, take $f \in End(S)$ as follows

$$f(x_\xi) := \begin{cases} f_\mu(x_\mu) & \text{if } \xi = \mu, \\ \varphi_{\mu,\nu}(f_\mu(y_\mu)) & \text{if } \xi = \nu \text{ and } \varphi_{\mu,\nu}(y_\mu) = x_\nu, \end{cases}$$

By hypothesis, there exists $f' \in End(S)$ such that $ff'f = f$ and $ff' = f'f$. In this case $f'|_{S_\mu} \in End(S_\mu)$. Thus $f_\mu f'_\mu f_\mu(x_\mu) = ff'f(x_\mu) = f(x_\mu) = f_\mu(x_\mu)$ and $f'_\mu f_\mu(x_\mu) = f_\mu f'_\mu(x_\mu)$ for each $x_\mu \in S_\mu$. This implies f_μ is completely regular and therefore $End(S_\mu)$ is completely regular. □

The converse is also true.

Lemma 6.5.3. *Let $Y = \{\mu, \nu\}$ with $\nu < \mu$ and let $S = [Y; S_\alpha, e_\alpha, \varphi_{\alpha,\beta}]$ be a non-trivial strong semilattice of left simple semigroups with surjective defining homomorphisms $\varphi_{\alpha,\beta}$. If the monoid $End(S_\xi)$ is completely regular for each $\xi \in Y$, then the monoid $End(S)$ is completely regular.*

Proof. Take $f \in End(S)$. Then $\underline{f} \in End(Y)$ which is completely regular. Then there exists $s \in End(Y)$ such that $\underline{f}s\underline{f} = \underline{f}$ and $\underline{f}s = s\underline{f}$. Let $x_\nu \in S_\nu$, there exists $y_\mu \in S_\mu$ such that $\varphi_{\mu,\nu}(y_\mu) = x_\nu$ since $\varphi_{\mu,\nu}$ is surjective.

Case 1. $\underline{f}(\mu) = \underline{f}(\nu) = \nu$, then we have $\nu = \underline{f}s(\nu)) = s(\underline{f}(\nu)) = s(\nu)$. Then $f_\nu \varphi_{\mu,\nu} = f_\mu$ where $f_\nu \in End(S_\nu)$ and $End(S_\nu)$ is completely regular, there exists $f'_\nu \in End(S_\nu)$ such that $f_\nu f'_\nu f_\nu = f_\nu$ and $f_\nu f'_\nu = f'_\nu f_\nu$. Using Construction 3.3.1, for every $x_\xi \in S,\ \xi \in Y$, take $f' \in End(S)$ as follows

$$f'(x_\xi) := \begin{cases} f'_\nu(x_\nu) & \text{if } \xi = \nu, \\ f'_\nu(\varphi_{\mu,\nu}(x_\mu)) & \text{if } \xi = \mu. \end{cases}$$

Then $ff'f = f$ and

$$\begin{aligned} ff'(x_\nu) &= f_\nu f'_\nu(x_\nu) \\ &= f'_\nu f_\nu(x_\nu) \ (End(S_\nu) \text{ is completely regular}) \\ &= f'f(x_\nu) \end{aligned}$$

and

$$\begin{aligned} f'f(x_\mu) &= f'(f_\nu(\varphi_{\mu,\nu}(x_\mu))) \\ &= f'_\nu(f_\nu(\varphi_{\mu,\nu}(x_\mu))) \\ &= f_\nu(f'_\nu(\varphi_{\mu,\nu}(x_\mu))) \ (End(S_\nu) \text{ is completely regular}) \\ &= ff'(x_\mu). \end{aligned}$$

Case 2. $\underline{f}(\mu) = \underline{f}(\nu) = \mu$, then $\mu = \underline{f}(s(\nu)) = s(\underline{f}(\nu)) = s(\mu)$. Then $f_\nu \varphi_{\mu,\nu} = f_\mu$ where $f_\mu \in End(S_\mu)$ and $End(S_\mu)$ is completely regular, there exists $f'_\mu \in End(S_\mu)$ such that $f_\mu f'_\mu f_\mu = f_\mu$ and $f_\mu f'_\mu = f'_\mu f_\mu$. Using Construction 6.1.1, for every $x_\xi \in S,\ \xi \in Y$, take $f' \in End(S)$ as follows

$$f'(x_\xi) := \begin{cases} f'_\mu(x_\mu) & \text{if } \xi = \mu, \\ \varphi_{\mu,\nu}(f'_\mu(y_\mu))) & \text{if } \xi = \nu \text{ and } \varphi_{\mu,\nu}(y_\mu) = x_\nu. \end{cases}$$

Then $ff'f = f$ and

$$\begin{aligned} ff'(x_\mu) &= f_\mu f'_\mu(x_\mu) \\ &= f'_\mu f_\mu(x_\mu) \ (End(S_\mu) \text{ is completely regular}) \\ &= f'f(x_\mu) \end{aligned}$$

and

$$\begin{aligned} ff'(x_\nu) &= f(\varphi_{\mu,\nu}(f'_\mu(y_\mu))) \ (\text{ where } \varphi_{\mu,\nu}(y_\mu) = x_\nu) \\ &= f_\nu(\varphi_{\mu,\nu}(f'_\mu(y_\mu))) \\ &= (f_\nu(\varphi_{\mu,\nu}))(f'_\mu(y_\mu)) \\ &= f_\mu(f'_\mu(y_\mu)) \\ &= f'_\mu(f_\mu(y_\mu)) \ (End(S_\mu) \text{ is completely regular}) \\ &= ff'_\nu(\varphi_{\mu,\nu}(y_\mu)) \\ &= ff'_\nu(x_\nu) \\ &= ff'(x_\nu). \end{aligned}$$

Case 3. $\underline{f}(\mu) = \mu$, $\underline{f}(\nu) = \nu$. Let $s \in End(Y)$ with $s(\nu) = \nu$, $s(\mu) = \mu$ such that $\underline{f}s = s\underline{f}$. Since $f_\mu \in End(S_\mu)$ and $End(S_\mu)$ is completely regular, there exists $f'_\mu \in End(S_\mu)$ such that $f_\mu f'_\mu f_\mu = f_\mu$ and $f_\mu f'_\mu = f'_\mu f_\mu$. Using Construction 6.1.1, for every $x_\xi \in S$, $\xi \in Y$, take $f' \in End(S)$ as follows

$$f'(x_\xi) := \begin{cases} f'_\mu(x_\mu) & \text{if } \xi = \mu, \\ \varphi_{\mu,\nu}(f'_\mu(y_\mu))) & \text{if } \xi = \nu \text{ and } \varphi_{\mu,\nu}(y_\mu) = x_\nu, \end{cases}$$

Then $ff'f = f$ and

$$\begin{aligned} ff'(x_\mu) &= f_\mu f'_\mu(x_\mu) \\ &= f'_\mu f_\mu(x_\mu) \ (End(S_\mu) \text{ is completely regular}) \\ &= f'f(x_\mu) \end{aligned}$$

and

$$\begin{aligned} f'f(x_\nu) &= f'(f_\nu(\varphi_{\mu,\nu}(y_\mu))) \ (\text{ where } \varphi_{\mu,\nu}(y_\mu) = x_\nu) \\ &= f'_\nu(\varphi_{\mu,\nu}(f_\mu(y_\mu)) \\ &= f'_\nu(\varphi_{\mu,\nu}(f_\mu(y_\mu))) \\ &= \varphi_{\mu,\nu}(f'_\mu f_\mu(y_\mu)) \\ &= \varphi_{\mu,\nu}(f_\mu f'_\mu(y_\mu)) \\ &= f_\nu(\varphi_{\mu,\nu}(f'_\mu(y_\mu))) \\ &= f_\nu f'_\nu(\varphi_{\mu,\nu}(y_\mu)) \\ &= ff'(x_\nu). \end{aligned}$$

Therefore f is completely regular. Hence $End(S)$ is completely regular. □

In the next theorem we get from Lemmas 6.5.2 and 6.5.3.

Theorem 6.5.1. *Let $S = [Y; S_\alpha, e_\alpha, \varphi_{\alpha,\beta}]$ be a non-trivial strong semilattice of left simple semigroups with surjective defining homomorphisms $\varphi_{\alpha,\beta}$. Then the monoid $End(S)$ is completely regular if and only if $|Y| = 2$ and the monoid $End(S_\xi)$ is completely regular for every $\xi \in Y$.*

6.6 Idempotent monoids

In this section we consider strong semilattices of left simple semigroups with surjective defining homomorphisms whose endomorphism monoids are idempotent.

Lemma 6.6.1. *Let $S = [Y; S_\alpha, e_\alpha, \varphi_{\alpha,\beta}]$ be a non-trivial strong semilattice of left simple semigroups with surjective defining homomorphisms $\varphi_{\alpha,\beta}$. If the monoid $End(S)$ is idempotent, then $|Y| = 2$ and the monoid $End(S_\xi)$ is idempotent for every $\xi \in Y$.*

Proof. By Lemma 3.2.3, the monoid $End(Y)$ is idempotent. This implies $|Y| = 2$ by Proposition 2.2.1. Assume $Y = \{\nu, \mu\}, \nu < \mu$. We next show that $End(S_\xi)$ is idempotent for every $\xi \in Y$.

Case 1. We show that $End(S_\nu)$ is idempotent, take $f_\nu \in End(S_\nu)$. Using Construction 3.3.1, for every $x_\xi \in S,\ \xi \in Y$, take $f \in End(S)$ as follows

$$f(x_\xi) = \begin{cases} f_\nu(x_\nu) & \text{if } \xi = \nu, \\ f_\nu(\varphi_{\mu,\nu}(x_\mu)) & \text{if } \xi = \mu. \end{cases}$$

By hypothesis f is idempotent. Thus $f_\nu f_\nu(x_\nu) = ff(x_\nu) = f(x_\nu) = f_\nu(x_\nu)$. This implies that f_ν is idempotent and therefore $End(S_\nu)$ is idempotent.

Case 2. We show that $End(S_\mu)$ is idempotent. Take $f_\mu \in End(S_\mu)$. Let $x_\nu \in S_\nu$, there exists $y_\mu \in S_\mu$ such that $\varphi_{\mu,\nu}(y_\mu) = x_\nu$ since $\varphi_{\alpha,\nu}$ is surjective. Let $f_\nu \in End(S_\nu)$ with $f_\nu(x_\nu) := \varphi_{\mu,\nu}(f_\mu(y_\mu))$ where $\varphi_{\mu,\nu}(y_\mu) = x_\nu$. Using Construction 6.1.1, for every $x_\xi \in S,\ \xi \in Y$, take $f \in End(S)$ as follows

$$f(x_\xi) = \begin{cases} f_\mu(x_\mu) & \text{if } \xi = \mu, \\ \varphi_{\mu,\nu}(f_\mu(y_\mu)) & \text{if } \xi = \nu \text{ where } \varphi_{\mu,\nu}(y_\mu) = x_\nu. \end{cases}$$

By hypothesis f is idempotent. Then

$$\begin{aligned} f_\mu(f_\mu(x_\mu)) &= ff(x_\mu) \\ &= f(x_\mu) \\ &= f_\mu(x_\mu). \end{aligned}$$

Thus f_μ is idempotent, and therefore $End(S_\mu)$ is idempotent. □

The converse is also true.

Lemma 6.6.2. *Let $Y = \{\nu, \mu\}$ with $\nu < \mu$ and let $S = [Y; S_\alpha, e_\alpha, \varphi_{\alpha,\beta}]$ be a non-trivial strong semilattice of left simple semigroups with surjective defining homomorphisms $\varphi_{\alpha,\beta}$. If the monoids $End(S_\mu)$ and $End(S_\nu)$ are idempotent, then the monoid $End(S)$ is idempotent.*

Proof. Take $f \in End(S)$. Then $\underline{f} \in End(Y)$ is a semilattice endomorphism on Y.

Case 1. $\underline{f}(\mu) = \underline{f}(\nu) = \nu$. Then

$$f_\nu(\varphi_{\mu,\nu}(x_\nu)) = \varphi_{\underline{f}(\mu),\underline{f}(\nu)}(f_\mu(x_\mu)) = f_\mu(x_\mu)$$

where $f_\mu \in Hom(S_\mu, S_\nu)$ and $f_\nu \in End(S_\nu)$ in this case. By hypothesis, $End(S_\nu)$ is idempotent, so that f_ν is idempotent.

We now consider

$$ff(x_\nu)) = f_\nu(f_\nu(x_\nu)) = f_\nu(x_\nu) = f(x_\nu)$$

and

$$ff(x_\mu)) = f(f_\mu(x_\mu)) = f_\nu(f_\nu(\varphi_{\mu,\nu}(x_\nu))) = f_\nu(\varphi_{\mu,\nu}(x_\nu)) = f_\mu(x_\mu) = f(x_\mu).$$

Thus f is idempotent.

Case 2. $\underline{f}(\mu) = \underline{f}(\nu) = \mu$. Then

$$f_\nu(\varphi_{\mu,\nu}(x_\nu)) = \varphi_{\underline{f}(\mu),\underline{f}(\nu)}(f_\mu(x_\mu)) = f_\mu(x_\mu)$$

where $f_\nu \in Hom(S_\nu, S_\mu)$ and $f_\mu \in End(S_\mu)$ in this case. $ff(x_\mu)) = f_\mu(f_\mu(x_\mu)) = f_\mu(x_\mu) = f(x_\mu)$ and

$$\begin{aligned} f(f(x_\nu)) &= f(f_\nu(x_\nu)) \\ &= f(f_\nu(\varphi_{\mu,\nu}(y_\mu))) \text{ where } \varphi_{\mu,\nu}(y_\mu) = x_\nu \\ &= f(f_\mu(y_\mu)) \\ &= f_\mu(f_\mu(y_\mu)) \\ &= f_\mu(y_\mu) \\ &= f_\nu(\varphi_{\mu,\nu}(y_\mu)) \\ &= f_\nu(x_\nu) \\ &= f(x_\nu). \end{aligned}$$

Thus f is idempotent.

Case 3. $\underline{f}(\mu) = \mu$, $\underline{f}(\nu) = \nu$. Then $f_\nu \in End(S_\nu)$ and $f_\mu \in End(S_\mu)$ with both are idempotents by hypothesis, and therefore f is idempotent. Thus $End(S)$ is idempotent. □

The following theorem follows directly from Lemmas 6.6.1 and 6.6.2.

Theorem 6.6.1. *Let $S = [Y; S_\alpha, e_\alpha, \varphi_{\alpha,\beta}]$ be a non-trivial strong semilattice of left simple semigroups with surjective defining homomorphisms $\varphi_{\alpha,\beta}$. Then the monoid $End(S)$ is idempotent if and only if $Y = \{\nu, \mu\}$ and the monoids $End(S_\nu)$ and $End(S_\mu)$ are idempotent for each $\xi \in Y$.*

Chapter 7

Arbitrary defining homomorphisms

Now we consider strong semilattices of semigroups in which the defining homomorphisms are arbitrary and $Y = \{\nu, \mu\},\ \nu < \mu$.

7.1 Regular monoids

If the defining homomorphisms $\varphi_{\alpha,\beta}$ are not isomorphisms or constant, then further complications arise. Take the following example (see also [4]), consider strong semilattices of semigroups with the two-element chain $Y = \{\nu, \mu\},\ \nu < \mu$. There are three endomorphisms of the chain $\nu < \mu$. They give three types of endomorphisms of S.

(1) $\underline{f}(\nu) = \underline{f}(\mu) = \nu$ then $f_\nu \in End(S_\nu)$ and $f_\mu \in Hom(S_\mu, S_\nu)$ such that $f_\mu(x_\mu) = f_\nu \varphi_{\mu,\nu}(x_\mu)$ for every $x_\mu \in S_\mu$,

(2) $\underline{f}(\nu) = \underline{f}(\mu) = \mu$ then $f_\nu \in Hom(S_\nu, S_\mu)$ and $f_\mu \in End(S_\mu)$ such that $f_\mu(x_\mu) = f_\nu \varphi_{\mu,\nu}(x_\mu)$ for every $x_\mu \in S_\mu$,

(3) $\underline{f}(\nu) = \nu$ and $\underline{f}(\mu) = \mu$ then $f_\nu \in End(S_\nu)$ and $f_\mu \in End(S_\mu)$ such that $\varphi_{\mu,\nu} f_\mu(x_\mu) = f_\nu \varphi_{\mu,\nu}(x_\mu)$ for every $x_\mu \in S_\mu$. We rewrite

$$\Theta = \{(f_\nu, f_\mu) \in End(S_\nu) \times End(S_\mu) \mid f_\nu \varphi_{\mu,\nu} = \varphi_{\mu,\nu} f_\mu\}.$$

It is clear that for $(f_\nu, f_\mu) \in \Theta$ then $f_\nu(Im(\varphi_{\mu,\nu})) \subseteq Im(\varphi_{\mu,\nu})$ and $f_\mu(Ker(\varphi_{\mu,\nu})) \subseteq Ker(\varphi_{\mu,\nu})$ where

$$Ker(\varphi_{\mu,\nu}) = \{(x, y) \in S_\mu \mid \varphi_{\mu,\nu}(x) = \varphi_{\mu,\nu}(y)\}.$$

If $\varphi_{\mu,\nu}$ is surjective, then the condition $f_\mu(Ker(\varphi_{\mu,\nu})) \subseteq Ker(\varphi_{\mu,\nu})$ implies that f_μ determines f_ν, so we simplify the description of Θ to

$$\Theta = \{f \in End(S_\mu) \mid f_\mu(Ker(\varphi_{\mu,\nu})) \subseteq Ker(\varphi_{\mu,\nu})\}.$$

If $\varphi_{\mu,\nu}$ is injective, then f_ν determines f_μ, so we simplify the description of Θ to

$$\Theta = \{f \in End(S_\nu) \mid f_\nu(Im(\varphi_{\mu,\nu})) \subseteq Im(\varphi_{\mu,\nu})\}.$$

Lemma 7.1.1. *Let* $Y = \{\nu, \mu\}$ *with* $\nu < \mu$ *and let* $S = [Y; S_\alpha, e_\alpha, \varphi_{\alpha,\beta}]$, $\varphi_{\mu,\nu} \neq c_{e_\nu}$ *be a non-trivial strong semilattice of semigroups with* $\nu = \wedge Y$. *Then* $End(S)$ *is regular if and only if the following conditions are satisfied:*

(R1) $End(S_\nu)$ *is regular,*

(R2) for every $f_\nu \in Hom(S_\nu, S_\mu)$, *there exists* $f'_\nu \in End(S_\nu)$ *such that* $f_\nu f'_\nu \varphi_{\mu,\nu} f_\nu = f_\nu$,

(R3) for every $(f_\nu, f_\mu) \in \Theta$, *there exists* $(f'_\nu, f'_\mu) \in \Theta$, *such that* $f_\nu f'_\nu f_\nu = f_\nu$, *and* $f_\mu f'_\mu f_\mu = f_\mu$.

Proof. Necessity. 1) Take $f_\nu \in End(S_\nu)$. Using Construction 3.3.1, for every $x_\xi \in S$, $\xi \in Y$, take $f \in End(S)$ as follows

$$f(x_\xi) := \begin{cases} f_\nu(x_\nu) & \text{if } \xi = \nu, \\ f_\nu(\varphi_{\mu,\nu}(x_\mu)) & \text{if } \xi = \mu. \end{cases}$$

By hypothesis there exists $f' \in End(S)$ such that $ff'f = f$. Then $f_\nu f'_\nu f'_\nu = f_\nu$ where $f'_\nu \in End(S_\nu)$, so that f_ν is regular and therefore $End(S_\nu)$ is regular.

2) Take $f_\nu \in Hom(S_\nu, S_\mu)$. Using Construction 3.3.1, for every $x_\xi \in S$, $\xi \in Y$, take $f \in End(S)$ as follows

$$f(x_\xi) := \begin{cases} f_\nu(x_\nu) \in S_\mu & \text{if } \xi = \nu, \\ f_\nu(\varphi_{\mu,\nu}(x_\mu)) & \text{if } \xi = \mu. \end{cases}$$

By hypothesis there exists $f' \in End(S)$ such that $ff'f = f$. In this case f' must be of the type (2), i.e., $f'_\mu \in Hom(S_\mu, S_\nu)$ and $f'_\nu \in End(S_\nu)$. Since $f' \in End(S)$ we have $f'_\nu(\varphi_{\mu,\nu}(x_\mu)) = f'_\mu(x_\mu)$. Therefore

$$f_\nu(x_\nu) = f_\nu f'_\mu f_\nu(x_\nu) = f_\nu f'_\nu \varphi_{\mu,\nu} f_\nu(x_\nu)$$

for every $x_\nu \in S$.

3) For every $(f_\nu, f_\mu) \in (End(S_\nu)) \times (End(S_\mu))$ with $f_\nu \varphi_{\mu,\nu} = \varphi_{\mu,\nu} f_\mu$. Define $f \in End(S)$ as follows $f(x_\nu) := f_\nu(x_\nu)$ and $f(x_\mu) := f_\mu(x_\mu)$ such that $f_\nu \varphi_{\mu,\nu} = \varphi_{\mu,\nu} f_\mu$. By hypothesis there exists $f' \in End(S)$ such that $ff'f = f$ and $f'_\nu \varphi_{\mu,\nu} = \varphi_{\mu,\nu} f'_\nu$ where $f'_\nu \in End(S_\nu)$, $f'_\mu \in End(S_\mu)$ such that $f_\nu f'_\nu f_\nu = f_\nu$, $f_\mu f'_\mu f_\mu = f_\mu$.

Sufficiency. Take $f \in End(S)$. Then $\underline{f} \in End(Y)$ consists of three types.

(1) If $\underline{f}(\nu) = \underline{f}(\mu) = \nu$ then $f_\nu \in End(S_\nu)$ and $f_\mu \in Hom(S_\mu, S_\nu)$ such that $f_\mu(x_\mu) = f_\nu\varphi_{\mu,\nu}(x_\mu)$ for every $x_\mu \in S_\mu$, by Condition 1) there exists $f'_\nu \in End(S_\nu)$ such that $f_\nu f'_\nu f_\nu = f_\nu$. Using Construction 3.3.1, for every $x_\xi \in S$, $\xi \in Y$, take $f' \in End(S)$ as follows

$$f'(x_\xi) := \begin{cases} f'_\nu(x_\nu) & \text{if } \xi = \nu, \\ f'_\nu(\varphi_{\mu,\nu}(x_\mu)) & \text{if } \xi = \mu. \end{cases}$$

(2) If $\underline{f}(\nu) = \underline{f}(\mu) = \mu$ then $f_\nu \in Hom(S_\nu, S_\mu)$ and $f_\mu \in End(S_\mu)$ such that $f_\mu(x_\mu) = f_\nu\varphi_{\mu,\nu}(x_\mu)$ for every $x_\mu \in S_\mu$, by Condition 2) there exist $f'_\nu \in End(S_\nu)$ such that $f_\nu = f_\nu f'_\nu \varphi_{\mu,\nu} f_\nu$. Using Construction 3.3.1, for every $x_\xi \in S$, $\xi \in Y$, take $f \in End(S)$ as follows

$$f'(x_\xi) := \begin{cases} f'_\nu(x_\nu) & \text{if } \xi = \nu, \\ f'_\nu(\varphi_{\mu,\nu}(x_\mu)) & \text{if } \xi = \mu. \end{cases}$$

(3) If $\underline{f}(\nu) = \nu$ and $\underline{f}(\mu) = \mu$, by Condition 3) we define $f' \in End(S)$ with $f'(x_\nu) = f'_\nu(x_\nu)$ and $f'(x_\mu) = f'_\mu(x_\mu)$ with $f'_\nu\varphi_{\mu,\nu} = \varphi_{\mu,\nu} f'_\mu$.

Thus we get $ff'f = f$, and therefore f is regular. □

Example 7.1.1. Consider a two-element semilattice $Y = \{\nu, \mu\}$, $\nu < \mu$. Let $G_\nu = G_\mu = \mathbb{Z}_6$ where $\mathbb{Z}_6$ is the additive groups modulo 6. Take the strong semilattice of groups $S = G_\nu \bigcup G_\mu$ with the defining homomorphisms as shown below.

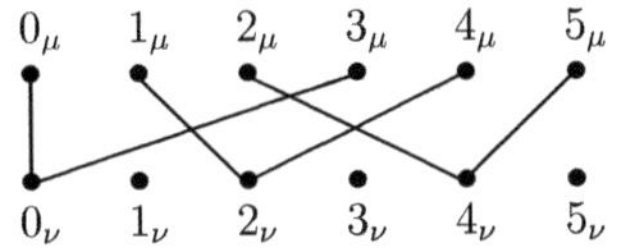

Take $f_\nu = \begin{pmatrix} 0_\nu & 1_\nu & 2_\nu & 3_\nu & 4_\nu & 5_\nu \\ 0_\mu & 3_\mu & 0_\mu & 3_\mu & 0_\mu & 3_\mu \end{pmatrix} \in Hom(G_\nu, G_\mu)$

Thus $\varphi_{\mu,\nu} f_\nu = \begin{pmatrix} 0_\nu & 1_\nu & 2_\nu & 3_\nu & 4_\nu & 5_\nu \\ 0_\nu & 0_\nu & 0_\nu & 0_\nu & 0_\nu & 0_\nu \end{pmatrix}$ does not exist $f'_\nu \in End(G_\nu)$ such that $f_\nu f'_\nu \varphi_{\mu,\nu} f_\nu = f_\nu$. This implies that $End(S)$ is not regular since there exists

$$f = \begin{pmatrix} 0_\nu & 1_\nu & 2_\nu & 3_\nu & 4_\nu & 5_\nu & 0_\mu & 1_\mu & 2_\mu & 3_\mu & 4_\mu & 5_\mu \\ 0_\mu & 3_\mu & 0_\mu & 3_\mu & 0_\mu & 3_\mu & 0_\mu & 0_\mu & 0_\mu & 0_\mu & 0_\mu & 0_\mu \end{pmatrix}$$

such that f has no an inverse element in $End(S)$.

Overview

The following table concludes all the results of this thesis. We observe that we do not know that for which groups G, the monoid $End(G)$ is left inverse or completely regular or other properties except for the monoid $End(G)$ is idempotent if and only if $G \in \{\mathbb{Z}_1, \mathbb{Z}_2\}$.

	$S=[Y;T_\alpha,e_\alpha,\varphi_{\alpha,\beta}]$ with bijective defining homomorphisms $\varphi_{\alpha,\beta}$			$S=[Y_{0,m};T_\alpha,e_\alpha,\varphi_{\alpha,\beta}]$ with surjective defining homomorphisms $\varphi_{\alpha,\beta}$
$End(S)$	T_α is a left simple semigroup	$T_\alpha=G_\alpha$ is a group	$T_\alpha=L_{n_\alpha}\times G_\alpha$ is a left group	T_α is a left simple semigroup
regular $\Leftrightarrow$	1) $End(Y)$ is regular 2) $End(T)$ is regular Theorem 3.2.3	1) $End(Y)$ is regular 2) $End(G)$ is regular Corollary 4.1.2, Corollary 5.1.8		$Hom(S_\alpha,S_\beta)$ is hom-regular Theorem 6.1.1
idempotent-closed $\Leftrightarrow$	1) $Y=Y_{0,m}$ 2) $End(T)$ is idempotent-closed Theorem 3.3.2	1) $Y=Y_{0,m}$ 2) $End(G)$ is idempotent-closed Corollary 4.2.2	1) $Y=Y_{0,m},\ n_\alpha=2,$ 2) $End(G)$ is idempotent-closed Corollary 5.2.2	$End(S_\alpha)$ is idempotent-closed Theorem 6.2.1
orthodox $\Leftrightarrow$	1) $Y=Y_{0,m}$ 2) $End(T)$ is orthodox Theorem 3.4.3	1) $Y=Y_{0,m},$ 2) $End(G)$ is orthodox Corollary 4.3.2	1) $Y=Y_{0,m},\ n_\alpha=2$ 2) $End(G)$ is orthodox Corollary 5.3.2	1) $End(S_\alpha)$ is idempotent-closed 2) $Hom(S_\alpha,S_\beta)$ is hom-regular Theorem 6.3.1
left inverse $\Leftrightarrow$	1) $Y=Y_{0,m},$ 2) $End(T)$ is left inverse Theorem 3.5.2	1)$Y=Y_{0,m},$ 2) $End(G)$ is left inverse Corollary 4.4.2	1) $Y=Y_{0,m},\ n_\alpha=2,$ 2) $End(G)$ is left inverse Corollary 5.4.2	$End(S_\alpha)$ is left inverse Theorem 6.4.1
completely regular $\Leftrightarrow$	1) $\|Y\|=2$ 2) $End(T)$ is completely regular Theorem 3.6.3	1) $Y=\{\nu,\mu\},$ 2) $End(G)$ is completely regular Corollary 4.5.2	1) $Y=\{\nu,\mu\},\ n_\alpha=2,$ 2) $End(G)$ is completely regular Corollary 5.5.2	1) $\|Y\|=2$ 2) $End(S_\alpha)$ is completely regular Theorem 6.5.1
idempotent $\Leftrightarrow$	1) $Y=\{\nu,\mu\}$ 2) $End(T)$ is idempotent Theorem 3.7.2	1) $Y=\{\nu,\mu\},$ 2) $G\in\{\mathbb{Z}_1,\mathbb{Z}_2\},\ G_\nu\neq G_\mu$ Corollary 4.6.2	1) $Y=\{\nu,\mu\},\ n_\alpha=1,$ 2) $G\in\{\mathbb{Z}_1,\mathbb{Z}_2\}$ Corollary 5.6.2	1) $\|Y\|=2$ 2) $End(S_\alpha)$ is idempotent Theorem 6.6.1

	$S = [Y; S_\alpha, e_\alpha, \varphi_{\alpha,\beta}]$ with constant defining homomorphisms and $\nu = \wedge Y$					
		S_α is a left simple semigroup		$S_\alpha = G_\alpha$ is a group		$S_\alpha = L_{n_\alpha} \times G_\alpha$ is a left group
regular	$\Rightarrow$	1) $End(Y)$ is regular, 2) $Hom(S_\nu, S_\alpha) = \{constant\}$, 3) $Hom(S_\alpha, S_\beta)$ is hom-regular Theorem 3.2.2	$\Rightarrow$	1) $End(Y)$ is regular, 2) $\lvert Hom(G_\nu, G_\alpha)\rvert = 1$, 3) $Hom(G_\alpha, G_\beta)$ is hom-regular Corollary 4.1.2	$\Rightarrow$	1) $End(Y)$ is regular, 2) $Hom(L_{n_\nu} \times G_\nu, L_{n_\alpha} \times G_\alpha) = \{\text{constant}\}$, 3) $Hom(G_\alpha, G_\beta)$ is hom-regular Corollary 5.1.6
	$\Leftarrow$	1) $Y = Y_{0,m}$, 2) $Hom(S_0, S_\alpha) = \{constant\}$, 3) $Hom(S_\alpha, S_\beta)$ is hom-regular, 4) S_0 contains one idempotent Theorem 3.2.1	$\Leftarrow$	1) $Y = Y_{0,m}$, 2) $\lvert Hom(G_0, G_\alpha)\rvert = 1$, 3) $Hom(G_\alpha, G_\beta)$ is hom-regular, Corollary 4.1.1	$\Leftarrow$	1) $Y = Y_{0,m}$, 2) $\lvert Hom(L_{n_0} \times G_0, L_\alpha \times G_\alpha)\rvert = 1$, 3) $Hom(G_\alpha, G_\beta)$ is hom-regular, 4) $\lvert L_{n_0}\rvert = 1$ Corollary 5.1.5
idempotent-closed $\Leftrightarrow$		1) $Y = Y_{0,m}$, 2) $End(S_\alpha)$ is idempotent-closed Theorem 3.3.1		1) $Y = Y_{0,m}$, 2) $End(G_\alpha)$ is idempotent-closed Corollary 4.2.1		1) $Y = Y_{0,m}$, $n_\alpha = 2$, 2) $End(G_\alpha)$ is idempotent-closed Corollary 5.2.1
orthodox	$\Rightarrow$	1) $Y = Y_{0,m}$, 2) $Hom(S_0, S_\alpha) = \{constant\}$, 3) $End(S_\alpha)$ is idempotent-closed 4) $Hom(S_\alpha, S_\beta)$ is hom-regular Theorem 3.4.1	$\Longleftrightarrow$	1) $Y = Y_{0,m}$, 2) $\lvert Hom(G_0, G_\alpha)\rvert = 1$, 3) $End(G_\alpha)$ is idempotent-closed 4) $Hom(G_\alpha, G_\beta)$ is hom-regular Corollary 4.3.1	$\Longleftrightarrow$	1) $Y = Y_{0,m}$, $n_\alpha = 2$, 2) $\lvert Hom(G_0, G_\alpha)\rvert = 1$, $n_0 = 1$ 3) $End(G_\alpha)$ is idempotent-closed 4) $Hom(G_\alpha, G_\beta)$ is hom-regular Corollary 5.3.1
	$\Leftarrow$	1) $Y = Y_{0,m}$, 2) $Hom(S_0, S_\alpha) = \{constant\}$, 3) $End(S_\alpha)$ is idempotent-closed 4) $Hom(S_\alpha, S_\beta)$ is hom-regular, 5) S_0 contains one idempotent Theorem 3.4.2				
left inverse $\Leftrightarrow$		1) $Y = Y_{0,m}$, 2) $End(S_\alpha)$ is left inverse Theorem 3.5.1		1)$Y = Y_{0,m}$, 2) $End(G_\alpha)$ is left inverse Corollary 4.4.1		1) $Y = Y_{0,m}$, $n_\alpha = 2$, 2) $End(G_\alpha)$ is left inverse Corollary 5.4.1
completely regular	$\Rightarrow$	1) $\lvert Y\rvert = 2$, 2) $Hom(S_\nu, S_\alpha) = \{constant\}$, 3) $End(S_\alpha)$ is completely regular Theorem 3.6.1	$\Longleftrightarrow$	1) $Y = \{\nu, \mu\}$, $\nu < \mu$, 2) $\lvert Hom(G_\nu, G_\mu)\rvert = 1$, 3) $End(G_\nu)$ $End(G_\mu)$ are completely regular Corollary 4.5.1	$\Longleftrightarrow$	1) $\lvert Y\rvert = 2$, $n_\alpha = 2$, 2) $\lvert Hom(G_\nu, G_\alpha)\rvert = 1$, $n_\nu = 1$ 3) $End(G_\alpha)$ is completely regular Corollary 5.5.1
	$\Leftarrow$	1) $\lvert Y\rvert = 2$, 2) $Hom(S_\nu, S_\alpha) = \{constant\}$, 3) $End(S_\alpha)$ is completely regular, 4) S_ν contains one idempotent Theorem 3.6.2				
idempotent $\Leftrightarrow$		1) $Y = \{\nu, \mu\}$, 2) $Hom(S_\nu, S_\mu) = \{constant\}$, 3) $End(S_\nu)$, $End(S_\mu)$ are idempotent Theorem 3.7.1		1) $Y = \{\nu, \mu\}$, 2) $G_\nu, G_\mu \in \{\mathbb{Z}_1, \mathbb{Z}_2\}$, $G_\nu \neq G_\mu$ Corollary 4.6.1		1) $Y = \{\nu, \mu\}$, $n_\alpha = 1$, 3) $G_\nu, G_\mu \in \{\mathbb{Z}_1, \mathbb{Z}_2\}$, $G_\nu \neq G_\mu$ Corollary 5.6.1

Bibliography

[1] Adams, M. E., S. Bulman-Fleming and M. Gould, *Endomorphism properties od algebraic structures*, Proc. Tennessee Topology Conf. (1996), World Scientific Pub. Co., 1997, NJ, 1-17.

[2] Adams, M. E. and M. Gould, *Finite semilattices whose monoids of endomorphisms are regular*, Transactions of the American Mathematical Society. **332**(1992), 647-665.

[3] Adams, M. E. and M. Gould, *Posets whose monoids of order-preserving maps are regular*, Order **6**(1989), 195-201. (Corrigendum, Order **7** (1990), 105.)

[4] Gilbert, Nick D. and Mohammad Samman *Clifford semigroups and seminear-rings of endomorphisms*, Journal of Algebra, **7**(2010), 110-119.

[5] Howie, J.M., *Fundamentals of Semigroup Theory*, Clarendon Press, Oxford 1995.

[6] Kasch, F. Adolf Mader *Regularity and substructures of Hom*, Communications in Algebra, **34**(2006), 1459-1478.

[7] Knauer, U. and M. Nieporte, *Endomorphisms of Graphs*, Arch. Math., Vol **52** (1989), 607-614.

[8] Knauer, U. and Worawiset, S., *Regular endomorphism monoids of Clifford Semigroups*, preprint.

[9] Krylov, P. A., A. V. Mikhalev and A. A. Tuganbaev, "Endomorphism Rings of Abelian Groups", Kluwer Academic Publishers, Dordrecht; Boston; London 2003.

[10] Mahmood, S.J., Meldrum, J.D.P and O'Carroll, L. *Inverse semigroups and near-rings*, J. London Math. Soc. (2), **23**(1981), 45-60.

[11] Meldrum, P. J. D., *Regular semigroups of endomorphisms of groups*, Recent Developments in the Algebraic, Analytical, and Topological Theory of Semigroups, Lecture Notes in Mathematics, **998**(1983), 374-384.

[12] Mitsch, H., *A Natural Partial Order For Semigroups*, proceedings of the american mathematical society, **97**(3)1986, 384-388.

[13] Petrich, M., *Inverse Semigroups*, J. Wiley, New York 1994.

[14] Petrich, M. and N. Reilly, "Completely Regular Semigroups", J. Wiley, New York 1999.

[15] Piotr A. Krylov, Alexander V. Mikhalev and Askar A. Tuganbaev, *Endomorphism Rings of Abelian Groups*, Kluwer Academic Publishers, Dordrecht/ Boston/London 2003.

[16] Puusemp, P., *Endomorphism semigroups of the generalized quaternion groups*, Acta et Comment. Univ. Tartuensis, **700**(1985), 42-49. (In Russian).

[17] Puusemp, P., *Idempotents of the endomorphism semigroups of groups*, Acta et Comment. Univ. Tartuensis, **366**(1975), 76-104. (In Russian).

[18] Puusemp, P., *On endomorphism semigroups of dihedral 2-groups and the alternating group A_4*, Acta et Comment. Univ. Tartuensis, **700**(1985), 76-104..

[19] Puusemp, P., *On the endomorphism semigroups of symmetric groups*, Acta et Comment. Univ. Tartuensis, **700**(1985), 76-104. (In Russian).

[20] Samman, M. and J. Meldrum, *On endomorphisms of semilattices of groups*, Algebra Colloquium, Vol. 12, **1**(2005), 93-100.

[21] Worawiset, S., *On endomorphisms of Clifford semigroups*, Semigroups, Acts and Categories with Applications to Graphs, Edited by V. Laan, S. Bulman-Fleming and R. Kaschek, Proc. Tartu (2007), 143-150.

Index

MIX
Papier aus verantwortungsvollen Quellen
Paper from responsible sources
FSC® C105338

Printed by Books on Demand GmbH, Norderstedt / Germany